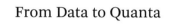

From Data to Quanta

NIELS BOHR'S
VISION OF PHYSICS

Slobodan Perović

FROM DATA TO QUANTA

The University of Chicago Press

Chicago and London

The University of Chicago Press, Chicago 60637
The University of Chicago Press, Ltd., London
© 2021 by The University of Chicago
Published 2021
Printed in the United States of America

30 29 28 27 26 25 24 23 22   21   1 2 3 4 5

ISBN-13: 978-0-226-79833-2 (cloth)
ISBN-13: 978-0-226-79847–9 (e-book)
DOI: https://doi.org/10.7208/chicago/9780226798479.001.0001

Library of Congress Cataloging-in-Publication Data

Names: Perović, Slobodan (Philosopher and historian of
    science), author.
Title: From data to quanta : Niels Bohr's vision of physics /
    Slobodan Perović.
Other titles: Neils Bohr's vision of physics
Description: Chicago ; London : The University of Chicago
    Press, 2021. | Includes bibliographical references and index.
Identifiers: LCCN 2021004008 | ISBN 9780226798332 (cloth) |
    ISBN 9780226798479 (ebook)
Subjects: LCSH: Bohr, Niels, 1885–1962. | Quantum theory—
    Philosophy. | Quantum theory—History.
Classification: LCC QC16.B63 P39 2021 | DDC 530.12—dc23
LC record available at https://lccn.loc.gov/2021004008

♾ This paper meets the requirements of ANSI/NISO Z39.48-
1992 (Permanence of Paper).

CONTENTS

1: INTRODUCTION

CONTROVERSY

It is puzzling that, despite Niels Bohr's exceptional prominence in the history of quantum physics and modern physics more generally, his physical concepts have often seemed unpolished, even misleading, to many physicists, historians, and philosophers of physics. As we will see shortly in more detail, this sort of attitude and Bohr's controversial reputation among these specialists is still surprisingly widespread. Resolving this puzzle motivated me in part to write this book, and understanding it helped me focus on some key aspects of Bohr's work. Although the comprehensive understanding of Bohr's work offered in the following pages should resolve the puzzle, the book far exceeds this initial motivation and sheds light on the work of one of the key figures of modern physics.

The book develops a novel approach to Bohr's understanding of physics and method of inquiry. My aim is an exploratory symbiosis of historical and philosophical analysis that uncovers the key aspects of Bohr's philosophical vision of physics within a given historical context. I argue that his vision was largely driven by his endeavor to develop a comprehensive perspective on novel experimental work, and his gradually developing accounts of the main features of experimentation. I will show that his distinctive research contributions were characterized by a multilayered or phased approach of building on basic experimental insights inductively in order to develop intermediary and then overarching (master) hypotheses. The strengths and limitations of this approach made him a thoroughly distinctive kind of physicist who ought to be investigated in a cross-disciplinary manner. I offer one such endeavor.

In my historical analysis, I focus mostly on Bohr's philosophical grasp of physics as it was driven by his practice during the early period of his work (roughly until the mid-1930s) when he developed his milestone contributions, while touching on a later period, substantially more removed from his actual "lab practice" and often addressed to a wider nonspecialist audience. The analysis highlights the crucial importance of experiment in his work, often in the form of "principles" based on experiment. It recognizes an important methodological coherence underlying Bohr's approach. This is ultimately more important in understanding his vision of physics than any definite philosophical stance we may try to identify in his work; Bohr's methodological concerns were prioritized throughout his career over more abstract epistemological and metaphysical concerns, as his famous debate with Albert Einstein demonstrates. When he explicitly addressed these further concerns, he did so by adding a final and, given the historical context, rather uncontroversial layer to his theoretical accounts. Finally, this methodological coherence is particularly apparent if viewed within the context of the community of quantum physicists who harbored distinctively different methodological approaches to quantum phenomena.

Philosophically, I develop an account of the relations between theory and experiment that prioritizes a semi-inductive (inductive-hypothetical) approach that shaped Bohr's practice. In such an approach we recognize stages or layers of hypotheses of different levels of generality, starting from the basic experimental ones. The basic experimental hypotheses defined by everyday language and notions of classical physics remain foundational; but what counts as a general master-level hypothesis harboring novel potentially nonclassical concepts is subject to reevaluation with new incoming experimental knowledge, which often ends up reducing it to a supporting role. In fact, throughout the course of his work, Bohr reassessed various features of both forms of hypotheses through the connection and increasing distinction asserted by the correspondence principle as an intermediary hypothesis during the formation of early quantum theory—and then, later on, during the formation of quantum mechanics, by the uncertainty principle.

When it comes to his main contributions, in short, Bohr's model of the atom introduced an ambitiously general theoretical approach synthesizing diverse research endeavors, while the correspondence principle was more speculative, and was intended as a forward-facing methodological device attempting to link rational theory and experimental work. During the development of quantum mechanics about a decade later, complementarity was developed as a synthetic perspective embedded in the experiments, primarily to explicate the limits and relationship between novel and diverse formal approaches and methods.

Yet before we fully unpack this account, let us return to the puzzle I sketched above, as an intriguing but a useful entry point to an understanding of Bohr's vision and practice of physics.

*

Bohr's peers considered him, "by tacit consent . . . the leader towards whom all turned for guidance and inspiration" (Rosenfeld in Bohr 1972–2008, vol. 1, xxxi). Yet some aspects of his work were vigorously criticized by some of the most prominent physicists of the era, such as Erwin Schrödinger and Einstein. Although the view of the microphysical world Bohr advocated in the mid- and late 1920s under the auspices of his principle of complementarity appears to have been closely tied to the experimental advances of his day, Schrödinger and Einstein were very reluctant to embrace it, seeing it as an obscure attempt to reconcile mutually contradictory concepts of particles and waves. And Bohr's precomplementarity correspondence principle concerning classical and quantum states (Bohr 1913a, 1922a), though central to the so-called old quantum theory, initially created a similar controversy, and was seen by some as embracing the essentially conflicting features of quantum and classical mechanics. Moreover, his breakthrough model of the atom (Bohr 1913a, 1913c, 1913d) was criticized throughout the period and, in fact, has been criticized ever since for discarding the spatial continuity of physical processes by introducing the "quantum jumps" of electrons from one discrete energy state (i.e., an orbit around the atomic nucleus) to another.

Yet these early criticisms pale in comparison to the more comprehensive ones developed by philosophers, historians, and physicists over the last several decades. For instance, Bohr's complementarity principle has been deemed an obscurantist account inherently open to diverse and mutually exclusive interpretations, and his approach to quantum phenomena has been judged an unprincipled imposition of his dogmatic metaphysical preferences on dissenters (e.g., Beller 1999, 1997, 1992; Bitbol 1996, 1995; Bub 1974).

Some authors have made determined efforts to debunk the principle of complementarity. James T. Cushing (1994) regards it as mostly empty rhetoric that operates through intellectual intimidation, while Jeffrey Bub (1974) argues that the complementarity principle is akin to Immanuel Kant's or Ludwig Wittgenstein's ultimately impenetrable philosophy. While Bohr's complementarity "endows an unacceptable theory of measurement with mystery and apparent profundity, where clarity would reveal an unsolved problem" (ibid., 46), "Bohr's contribution to the Copenhagen Interpretation" is "that of a remarkably successful propagandist" (ibid., 45). This harsh ideologue in the quantum physics community saw "the statistical relations of quantum mechanics as the confirmation of an approach to the problem of knowledge that had fascinated him since youth" (ibid.), and imposed it on others.[1] For his part, Imre Lakatos (1970) saw the continuous development of Bohr's model of the atom as a prime example of a degenerating research program; others followed suit, deeming the model inherently inconsistent (Jammer 1966).[2]

The physicist John Bell (2001, 197), well known for his foundational theorem in quantum mechanics, says:

> Rather than being disturbed by ambiguity in principle . . . Bohr seemed to take satisfaction in it. He seemed to revel in contradictions, for example between "wave" and "particle," that seem to appear in any attempt to go beyond the pragmatic level. Not to resolve these contradictions and ambiguities, but rather to reconcile us to them, he put forward a philosophy, which he called "complementarity."

Harsh words indeed, from arguably the most important figure in the post–World War II physics of the foundations of quantum mechanics.

The litany of complaints continues to this day. More recently, the harsh criticism reached more popular venues like *Forbes* magazine, where a prominent science journalist, Chad Orzel, (2015) stated, "Bohr is a pretty bad example of philosophy in physics, as he was maddeningly vague and a horribly unclear writer." His harsh assessment of Bohr's work and his impact on physics is aligned with that of Tim Maudlin (2018): "The obscurantism of Bohr and Heisenberg, which became known first as the 'Copenhagen Interpretation' and in its later incarnations as 'Shut Up and Calculate,' is a self-conscious abdication of the aim of physics, namely, to understand the nature of the physical world." In his latest, introductory book on quantum mechanics (Maudlin 2019), Maudlin deliberately omits Bohr's work and the Copenhagen Interpretation, as it does not meet the standards of a physical theory that "clearly and forthrightly address[es] two fundamental questions: what there is, and what it does" with "sharp mathematical description" and "dynamics . . . by precise equations describing how ontology will evolve" (ibid., xi). Instead of informing the reader of the details of Bohr's work, Maudlin refers his reader to previous unforgiving criticisms of it, as "our time is better spent presenting what is clear than decrying what is obscure" (ibid.).

Similar sentiments are echoed in a historical analysis aimed at a wider audience; Adam Becker's 2018 book provides an intriguing counterfactual history of quantum mechanics based in part on these sentiments.[3] And we should mention a prominent philosopher of science, namely David Albert, who recently offered an acerbic, albeit entertaining, version of an attitude found in the philosophy of the physics community, in a podcast series run by the outspoken theoretical physicist Sean Carroll.[4]

This sort of criticism and the impression it generates must be set against Bohr's stature among his contemporaries, if we agree that it is "a distortion to see Bohr's views as basically stemming from an a priori philosophical background" (Dieks 2017, 307). After his first big breakthrough on the model of the atom in 1913, Bohr quickly assumed a central role in the quantum physics community. Paul Ehrenfest's comment to a young physicist in 1929 convincingly attests to his status: "Now you are going to get to know Niels Bohr and that is the most important thing to happen in the life of a young physicist" (Casimir 1968, 109).

What happened to Bohr's reputation? Why are we now getting such a different reaction to his work? And how is this relevant to our understanding of his physics and its methods?

The skeptical attitudes to Bohr's work are diverse—voiced by specialists on the history and philosophy of quantum mechanics, various philosophers of physics, philosophers of science, physicists, science journalists, and so on—but many are an unfortunate consequence of the fact that both critics and defenders (e.g., Howard 2007; Landsman 2006; Chevalley 1994; Faye 1991; Murdoch 1987) too often focus on the search for an exact metaphysical or epistemological account to which they think Bohr may have subscribed and which, in turn, may have shaped his major contributions to physics. I should note that, in general, just as harsh criticism often deflects the analysis from what I consider the central aspect of Bohr's work—his methodology—so too does the enthusiasm of certain philosophers who rush to ascribe to Bohr certain metaphysical or epistemological views. Bohr's vision of physics has been identified as Kantian (Bitbol 2017; Cuffaro 2010; Chevalley 1994; Kaiser 1992), a transcendental research program (Bitbol 2017), an account of relational holism (Dorato 2017), or an account ripe with semantic and metasemantic implications (Osnaghi 2017). Parallels have recently been drawn between his work and Pragmatism in philosophy (Faye 2017; Heilbron 2013, 33), and even between his ontological views and the religious views of Søren Kierkegaard and those in the Talmud (Clark 2014). Also, over the last few decades many authors have concentrated on side issues in Bohr's work, focusing, for example, on his free-thinking mature phase, which often veered from a concrete experimental context but is philosophically intriguing. Moreover, leading physicists both sympathetic to and skeptical of Bohr's work (typically his later work) have insisted on Bohr's wholly spontaneous, as it were, and purely intuitive process of generating ideas, sometimes calling it a divine glance (Kramers 1935, 90).

Analyses of this sort are often valuable for understanding Bohr's overall vision of the physical world and related philosophical issues if we approach them cautiously. David Favrholdt (1992), by contrast, has articulated an opposing and perhaps exaggerated view, saying that since Bohr never studied philosophy systematically, his contributions stem from his physics alone (Heilbron 2013, 33). As we will see in due course, however, it is not always easy or desirable to try and disentangle physics from philosophical concerns, particularly epistemological and methodological concerns, especially when studying the emergence of a completely new theory. Yet there is something more fundamental at the core of Bohr's approach to physics that many philosophical interpreters and critics have missed—something other than what his occasional intellectual flourishes, musings, and wide-ranging elaborations of microphysical phenomena suggest if they are read apart from his practice of physics. Philosophers have often focused on these, but they are often a sideshow, especially when isolated from the rest of his work. The story of Bohr's method and the vision behind it may be disappointingly conventional in comparison.

In that sense, my account of Bohr's work is deflationary. A number of important philosophical influences on various aspects of, and stages, in Bohr's work can certainly be identified,[5] but the core of what Bohr did as a physicist is, generally speaking, a pretty standard experiment-driven inductive approach, a particular strand that he perfected and improved in the context of founding quantum theory and quantum mechanics. And as Dennis Dieks (2017, 303) recently emphasized, Bohr's works read very differently from "the tradition of foundational work that started in the early 1950s." One remarkable aspect of his approach is that he consciously and uncompromisingly stuck to it wherever it led him, including renouncing some of the key aspects of what came to be recognized as classical physics, as well as the principles other prominent physicists deemed inescapable, even though he was fully aware of potential losses entailed by such moves.

Instead of taking the metaphysical and epistemological background as the starting point for an analysis of Bohr's work, I undertake a historically sensitive philosophical analysis of the method that produced his breakthrough results. We primarily need to understand Bohr's method, the exact role Bohr played in the physics community, and the reasons for both. Only then can we properly judge his accomplishments and end the controversy. I also expect the story about Bohr's vision of physics to clarify for philosophers of science interested in methodological questions how exactly the repairs on the sailing ship of science were made, to use a well-worn phrase, during this major episode in the history of science, and thus to enrich their current inquires (e.g. what constitutes the basic experimental level of the inductive process, how the much discussed notion of "bridge laws" relates to the intermediary hypotheses crucial for the inductive process pursued by Bohr and others, the nature of modelling during the crafting of the quantum theory, and so on).

As the above brief summary suggests, there are two ways of approaching Bohr's work. First, we can attempt to understand his accounts of the physical world with the help of known and well-developed philosophical terms and doctrines. This doctrinal way is often favored by philosophers' attention. Second, we can try to understand his underlying approach to physical states and processes, and establish whether certain unifying principles or heuristics underlie it. The latter methodological kind of analysis can be pursued fairly independently of the former,[6] yet in my view it should ground the former, not the other way around.

The main danger of the first way is in reading preferred philosophical terms and doctrines into Bohr's account while distorting or even ignoring the methodological understanding. Moreover, the debates among physicists and their underlying epistemological and ontological views were in flux when Bohr produced his most important results. The physicists understandably kept crafting their positions, giving them up, changing them. Typically, only rough and barely adequate distinctions can be identified within such evolving understandings when assessed by the usual conceptual machinery of contemporary philosophy, despite our best efforts to "nail them down." In Bohr's case, however, the key general methodological understanding—the understanding of how exactly one ought to craft physical theory—remained stable, in contrast to, for example, Arnold Sommerfeld's methodology, which substantially changed between different periods of his work (Seth 2010). It is possible, of course, that there are no underlying unifying principles or heuristics and no coherent approach to physical states and process throughout Bohr's work, and that the doctrinal understanding may be the only way to understand him. But my analysis suggests otherwise.

Bohr's later elaborate metaphysical and epistemological considerations were typically derivative of his scientific practice and his reflections on the immediate method that shaped it. Yet sometimes, especially in his mature phase, they were fairly removed from it and targeted a wider nonspecialist audience. Therefore, it is crucial to understand properly the connection between Bohr's development of his epistemological attitude to scientific research, the metaphysical ideas he occasionally commented on, and the actual method of finding out about microphysical processes he gradually developed and excelled in. In fact, for most of his active career the methodological concerns and their tangible epistemic ramifications that were close to the experimental context were much more prominent and fundamental than the abstract epistemological and ontological considerations in his reflections. Bohr's work is certainly more "indicative of a physicist's rather than a philosopher's attitude" (Dieks 2017, 307) if we are thinking of a professional "philosopher" of the last hundred years or so. Yet Bohr's attitude was shared by only some members of the physics community at the time; as we will see in due course, others pursued physics in a much more philosophical (in the above sense) or mathematically driven manner.

BOHR'S VISION AND PRACTICE OF PHYSICS

The primary goal of this book is to trace the structure of Bohr's approach to physics by drawing on numerous historical studies. Its secondary goal is to assess critically a broad array of views of Bohr, which are often critical and often, as we have seen, portray him as a physicist who bullishly pursued his prior, quite contentious metaphysical and epistemological views.

The main goal of this book, that of advancing a methodological understanding of Bohr's work, is pursued through a historically motivated philosophical assessment of the scientific method as a constant and vigorous interplay between theory and experiment in the early development of quantum physics. Generally speaking, it is a case study that identifies the crucial traits of the experimentalist strand in modern physics prior to its transition to the industrial-scale science of the second half of the twentieth century, on the one hand, and to mathematically driven abstract theory, on the other. In particular, it is a philosophical study of Bohr's methodology, his way of doing physics, and his own reflections on his method. This book offers an account of the process characterizing Bohr's work in the context of a wider community of physicists. It is not primarily a historical study, though it draws heavily on the original texts, accounts of Bohr's contemporaries, and a number of historical studies.

I suggest that we should primarily consider the vision of physics Bohr explicated as a reflection on the way he and much of the community of quantum physicists practiced physics. We should do so by focusing on its key aspects in practice; crucially, his vision of physics was inherently related to the experiments as they were performed and discussed at the time. In general, Bohr's approach to physics, what he preached and practiced, stands in sharp contrast to the metaphysically and mathematically oriented approaches of some other physicists.

Bohr's most productive early work on the atomic structure and quantum mechanics that sprang from the experimental context is also the most interesting in his long oeuvre from the point of view of methodology and epistemology of physics. His early breakthrough contributions of the model of the atom and the correspondence principle resulted from this approach. Simply stated, Bohr attempted to grasp fully the entire body of potentially relevant experimental results in the form of novel intermediary and master hypotheses devised at different levels of interfacing experiments and theory. Contributing later to the emergence of quantum mechanics, he devised a provisional semi-inductive synthesis of various novel contending accounts of quantum phenomena and respective novel formalisms, in the shape of his complementarity principle which charted their respective limitations. The complementarity principle did not arrive in as forceful and unforeseen a manner as his earlier contributions, but it exerted a lasting methodological influence on the experimentalists in quantum mechanics.

Once we see that Bohr played the typical role of a laboratory mediator and excelled in the inductive-hypothetical process this required, we can understand better the way his work was generated, its role in developing novel concepts, its true limitations, and the adherence to and use of the complementarity approach by contemporary experimentalists, as opposed to theoreticians and philosophers far removed from the lab. This interpretive template should clarify Bohr's statements by setting them in the context of his entire oeuvre and its overarching aims.

It may be granted that the results Bohr's vision of physics produced—namely, his model of the atom, the principle of correspondence, and the complementarity principle—are philosophically wanting when we try to understand them from the point of view of the well-defined and well-developed theoretical and philosophical accounts of the physical world provided by academic philosophers. But they are also inescapably the place to look if we want to understand the method that produced such major breakthroughs in understanding quantum phenomena. And perhaps substantially different trajectories in developing quantum mechanics, much more in line with the preferences of contemporary philosophers of science critical of Bohr's work, were possible. Even so, it does not mean

Bohr's trajectory was not a reasonable experimentally led road to quantum mechanics.

*

Part 1 of this book explains the nature of Bohr's engagement with the experimental side of physics, and lays out the basic distinctions and concepts characterizing his approach. It explains how exactly Bohr saw the basic structure of the scientific method, especially the two-stage relationship between theory and experiment, and why this was at the heart of his much-discussed distinction between the classical and quantum concepts.

Experimental results are first delivered in the shape of reports on particular features of experiments, in the form of lower-level experimental hypotheses. Such features are characterized within the bounds of our everyday observational experience and the observational concepts suggested by it, with the help of classical physical concepts devised and solidified in previous work—a point Bohr continually emphasized. After the first stage of data collection, reports on experimental features and often vigorous debates on adequate lower-level hypotheses, we pass to the second and theoretical stage of the inductive process. In it, physicists try to make sense of diverse lower-level hypotheses by devising intermediary (supporting as well as constructive) hypotheses, and finally general master hypotheses that can satisfactorily account for them. These higher-level hypotheses can harbor nonclassical quantum concepts.

In part 1 of this book, I do not expect to convince the reader of my view of Bohr's approach to physics and its connection with the actual practice of physics from which it stemmed. Rather, I expect the reader to grasp the basic picture of that approach and Bohr's motivation for pursuing it. The following chapters discuss how Bohr's approach worked in practice, by drawing on historical scholarship. If the reader expects to fully understand Bohr's method purely conceptually without diving into the detailed history of his work, they will be deeply troubled with the rest of the book!

Part 2 offers a developed interpretation of Bohr's method, embedded in the detailed historical analyses and relevant examples concerning his model of the atom and his correspondence principle, as well as a discussion of their reception and understanding within the community of physicists working on quantum theory. The two-stage approach to physical phenomena led Bohr, even in his PhD dissertation, to question the received physical principles, irrespective of how compelling or inescapable they might have seemed, that did not agree with *the experimental context*—the relevant experimental results and the key features of the experiments that produced them. Sticking to this attitude, Bohr derived his model of the atomic structure as a master hypothesis via three supportive intermediary hypotheses—Ernest Rutherford's model of the atom, Max Planck's second quantum law of radiation, and the hypothesis of atom's stability—each subsuming particular experimental lower-level hypotheses stemming from a score of experiments with various, often interrelated aspects of atomic spectra, radiation, radioactivity, and electromagnetism. The model combined their different aspects into an analogy between the structure of the atom and a classical planetary system of electrons orbiting the nucleus, but only along the orbits permitted by quantum rules. The resulting model was a provisional master hypothesis induced from the experimental results and, as such, continuously refined in light of new experiments. And the correspondence principle stating, roughly, that the frequency of electrons transitioning between "higher" orbits was analogous to the classical mechanical account of electron frequencies was another key (constructive and versatile) intermediary hypothesis, a tool that bridged the quantum model and the classical records of the experimental results. The model's comprehensive grasp of lower-level hypotheses, especially the quantitative agreement with the spectral laws hitherto not considered relevant to the atomic structure, made it acceptable despite its conceptual deficiencies, of which Bohr himself was well aware.

Philosophers have occasionally taken up certain aspects of Bohr's model of the atom, especially the methodological and epistemological aspects of its construction (e.g., Weinert 2001; Norton 2000; Achinstein 1993; Hutten 1956). But the complementarity account sparked more interest among them and often, in some of the works already mentioned, was seen as emblematic of the alleged obscurity of Bohr's work. This may not be surprising, given that the model was a key element of the short-lived old quantum theory while complementarity addressed quantum mechanics, the key ingredient of contemporary physics. Yet, as we will see, the complementarity principle was a later result of the same sort of approach that produced the atomic model, and both were assigned the same provisional role by Bohr.

In the last section of part 2, I explain that Bohr's role as a moderator in the community of physicists working on quantum theory was inherently related to the nature of his approach to physical phenomena. I also discuss some previous attempts to explain how his model of the atom was generated, including the philosophical implications of Bohr's methodological views.

Part 3 focuses on the emergence of Bohr's principle of complementarity from the novel experimental context of the first half of the 1920s, and explains the role it played in crafting quantum mechanics. The complementarity account of quantum phenomena was Bohr's third major result of his two-stage inductive method, and the most controversial. It was an early, general, and experimentally adequate albeit provisional account of microphysical phenomena, aimed at reconciling novel and diverse experimental and formal results. In the words of Werner Heisenberg, in inducing it, Bohr was committed first and foremost to "the requirement of doing justice at the same time to the different experimental facts which find expression in the corpuscular theory on the one hand and the wave theory on the other" (Heisenberg in Bohr 1972–2008, vol. 6, 20–21). Two supporting hypotheses (wave mechanical and quantum-corpuscular) and a new constructive hypothesis Bohr co-crafted with Heisenberg (the uncertainty principle) were gradually devised from novel experiments, primarily those on wave interference and on the scattering of matter and light. A decisive step in establishing a new master hypothesis was a domain-specific equivalence (the domain defined by particular features of Bohr's atom) of two novel formal approaches (wave-mechanical and matrix-mechanical) that accounted for the two supporting hypotheses. The final result did not resonate with the metaphysically motivated customary intuitions of either wave-mechanical or quantum-corpuscular interpretations—such experimentalist-minded general accounts rarely do. Accordingly, understanding complementarity outside the context in which Bohr devised it can be detrimental to understanding it as a theoretical framework.

Part 4 points out that the Copenhagen interpretation of quantum mechanics was too crude to take on board all the subtleties of Bohr's work that were crucial for developing quantum mechanics—and understandably so, given the aim of the interpretation. I discuss Bohr's response to the famous Einstein-Podolsky-Rosen criticism of quantum mechanics, illuminating the clash between Bohr's approach, firmly grounded in the existing experimental context, and Einstein's prescient philosophical concerns. I conclude by explaining that Bohr's approach lost its primacy once the mathematical crafting of quantum mechanics and its derivatives, such as quantum electrodynamics and quantum field theory, became the focus of the physics community, though the features of his methodology have remained a mainstay in experimental quantum mechanics.

PHILOSOPHICAL AND HISTORICAL ANALYSIS

A reader keen to jump to the details of Bohr's work may wish to move to the next chapter now and revisit this section after reading the rest of the book. This section is intended primarily for philosophers of science and philosophers of physics; it situates the book and its aims within the larger context of philosophical and historical studies of science, albeit in a rather condensed form. That said, most points mentioned here will reappear throughout the book.

I have previously developed several arguments concerning Bohr's work in the mid-1920s and his principle of complementarity in particular (Perović 2017, 2013, 2008, 2006, 2005), and these comprise an important backdrop to the central argument of this book—namely, that there was a bottom-up inductive process at work, firmly entrenched in the relevant experimental context. I have previously identified the elements of this process in Bohr's approach to physics, but here I offer both a more comprehensive analysis of Bohr's method and his work, and a much more detailed characterization of the inductive process in the experiment-driven strand in physics that led to Bohr's key discoveries.[7]

The strong undercurrent of Baconian-style experimentalism and induction in Bohr's approach to physical phenomena belongs to a long experimentalist tradition within the complex human activity we call the scientific method. Like Francis Bacon and many others after him, Bohr emphasized the two-stage nature of the experimental process, in which the first experimental stage is both the foundation of and a tool for crafting the theory.[8] Philosophers have systematically downplayed the role of experiments in Bohr's work. This is true of earlier accounts of Bohr's work put forward by Thomas Kuhn (1987), Paul Feyerabend (1969), and Lakatos (1970),[9] and more recent accounts rarely focus on it. Lakatos even claimed that experiments played practically no role in establishing early quantum theory and could have been abolished for all practical purposes. Nor, according to him, did they play a decisive role in establishing quantum mechanics later on. This is a crucial misunderstanding, and as such it can provide only a marginally satisfying understanding of Bohr's work. One of the key goals of this book is to show that experiments played a pivotal role in developing all three of Bohr's main contributions, and to explain how they did so. The scale of Lakatos's misunderstanding shows how philosophers are too often prone to select the narrative of the history of science they deem relevant simply to accommodate their preferred philosophical views; this is emblematic of a wider disconnect between history and philosophy of science. Only recently has a recognizable interest in the experiments as fundamental to Bohr's methodology emerged (Camilleri 2017; Camilleri and Schlossauer 2015; Perovic 2013; Howard 1994). Previously, they were almost exclusively discussed with respect to the epistemic and conceptual framework within which Bohr understood quantum phenomena—in particular, the entanglement of the observer and the observed quantum phenomena—though perceptive historians have pointed out their crucial role in Bohr's work (Kragh 2012).

Throughout his career, Bohr was constantly in conversation with the experimentalists, the physicists interested in inventing and building experimental apparatus to test new and intriguing questions about physical phenomena. In fact, the key insights that led to the turning points in Bohr's work on the model of the atom, the correspondence principle, and complementarity sprang directly from these consultations.

There was a unique symbiosis of experiment and theory in the laboratories of Joseph John Thomson and Rutherford, where Bohr started his career. Theoretical physics had begun to be institutionally treated as a distinct subdiscipline at the beginning of the twentieth century (Seth 2010, sections 1 and 4), and some physicists were inclined to do physics in that vein alone, including some whose work was instrumental in developing quantum mechanics. Wilhelm Wien stated in 1890 that "theoretical physics today finds no takers" (ibid., 4). The figure of an ambitious theoretical physicist who rarely enters the lab and who focuses on developing impressive coherent mathematical models that can veer far from the experimental work became fully institutionally established only in the second half of the twentieth century. It is also not a coincidence that Schrödinger and Einstein, Bohr's most vigorous critics, were among the emerging figures of that sort. Measured by this standard, Bohr's work may seem inferior, or outright unacceptable.

The phase change in the development of modern physics and the turn to high-level abstract mathematically driven theory arguably started with Paul Dirac's work on symmetry, and on the building of quantum electrodynamics and then quantum field theory. Yet only after experiments bore results and stabilized the fundamentals of the theory was it sensible to turn to theoretical refinements or to work out various alternative interpretations, including those of David Bohm or Hugh Everett III.[10] Bohr was a central figure of the first phase.

We could label the process of generating hypotheses all the way up from the experimental work—the process Bohr practiced and reflected on—the inductive-hypothetical process. Although experimental hypotheses are generated from the accounts of particular experimental aspects, they are indeed all hypothetical in nature, starting with the basic experimental accounts. They are warranted by experimental techniques and reasonable experimental strategies that can be different in different laboratories, and these independently generated hypotheses subsequently provide the ground for generating more general theoretical hypotheses.

Some might see the recently well-explored notion of a scientific model as more adequate than the notion of hypothesis for the purposes of the present analysis. Yet the notion of hypothesis is more apt, as the notion of the model was already in use among quantum physicists at the beginning of the twentieth century, primarily as a label for general theoretical models (I label them master hypotheses), such as Bohr's model of the atom. In any case, the use of the notion at the time does not map well onto the current focus among philosophers of science, so the use of a more generic notion of hypothesis will avoid possible confusion. As I will explain in more detail in chapter 3, my notion of hypothesis includes the notions of principles, postulates, models, axioms, theoretical models, and theories—all used by physicists in various contexts across levels of the scientific process, and all having a broadly hypothetical character.

In essence, the lower-level experimental hypotheses are induced on the basis of a chain of experimental particulars. Any existing models and principles are of secondary importance; they are reassessed, used partially, or abandoned if necessary. The formation of experimental lower hypotheses is typically independent of the formation of hypotheses at the theoretical stage. Yet, as will become apparent especially in the historical account of the experimental work in spectroscopy, the experimental hypotheses regarding the same phenomenon can substantially differ when they are pursued by different experimentalists in different laboratories. And deliberations assign different weights to different experimental particulars. In other words, the experimental hypotheses that led to quantum theory and quantum mechanics had a relatively independent history; in that sense, they had a "life of their own," to use Ian Hacking's well-known phrase. But there were no immediate, unique, and widely agreed-upon interpretations of facts to be assimilated into the higher-level theoretical hypotheses. Quite the opposite, in fact; experimentalists often differed in their understanding of the relevance of various experimental details, and thus in their formulation of the lower hypotheses. The negotiating process involved in establishing an experimental hypothesis to be included in further, more abstract theoretical hypotheses can be long and arduous, as in the development of spectrometry and Bohr's model of the atom; or it can be less protracted, as in the case of the scattering experiments in the mid-1920s that laid the groundwork for quantum mechanics.

A recent debate among empiricists in philosophy of science makes a similar point about the interplay between observations and data.[11] Elisabeth Lloyd (2012) contrasts the "complex empiricism" that emphasizes precisely this interplay with the "direct empiricism" that adheres to a traditional emphasis on theories confirmed by the data. She analyzes the case of climate scientists who can be divided by adherence to these two epistemological categories. This distinction may be apt in the case of development of quantum mechanics as well. It was particularly true in the development of lower-level hypothesis, during the advance of old quantum theory and quantum mechanics, that "data are never naked, and measurement does not occur without the imposition of framing or generating theories and models" (Lloyd 2012, 392). And Bohr's empiricism was certainly much more "complex" than "direct." An important difference from Lloyd's case, however, was the extent to which this interplay between observations and data in the experimental work was independent from the high-level theory and confined to the lower-level hypotheses. The intermediary and higher-level hypotheses were typically introduced after the lower-level ones were fairly established.

Bohr insisted on the independence of experimental hypotheses and the results they elicited, and argued that more general hypotheses should be built primarily on an extensive understanding of them. This relative independence of experimental hypotheses was also reflected in the terminology and nature of concepts devised in the process (thus, quantum notions being confined to formulations of higher-level hypotheses, and experimental hypotheses formulated with everyday notions amended with the notions of classical physics).

But this was not the only approach. In general, to advance a higher-level hypothesis without close agreement with an already set experimental hypothesis following a shorter or more protracted debate, the scientist had essentially to devise their own novel experimental hypothesis from the existing experimental particulars and data. Thus, for instance, in the mid-1920s Schrödinger had to reinterpret the widely agreed-upon conclusions drawn from the scattering experiments if he was to argue in favor of his wave-mechanical interpretation. In yet another contrast to Bohr's approach, in one of the phases of his work, Sommerfeld tried a more direct reading of theoretical conclusions from experimental data. Finally, Heisenberg and other mathematically oriented physicists treated elicited experimental results as a starting point from which to develop an abstract mathematical model, without gradually shaping their novel hypotheses via experimental ones in the way that Bohr did (their approach being more akin to Lloyd's "direct empiricist" attitude). In other words, experiments played for Bohr a much more pronounced role in every stage of building the theoretical structure than they did for Heisenberg, Sommerfeld, or Schrödinger.

Thus, different physicists had different visions of how exactly and at what point the more general higher-level hypotheses connected with the lower ones—or, in other words, how and at what point the experimental results should be used. Were those results a tool to devise a theory at multiple stages or a statistical set with which to compare a mathematical model? Although Sommerfeld changed his approach later on, he initially practiced physics that drew on the experimental results in a manner analogous to solving particular problems in engineering. Mathematical solutions were applied primarily as tools of prediction and retrodiction by Heisenberg, Dirac, and Wolfgang Pauli, while physicists like Planck and Einstein ultimately sought to "subsume all physical phenomena under a few abstracted, generalized axioms" (Seth 2010, 2). Each of these approaches exhibited both advantages and disadvantages in various contexts. Bohr was deeply aware of the trade-offs between conceptual clarity and the heuristic value of the products of his own bottom-up approach to physics.

Bohr, Sommerfeld, Schrödinger, Heisenberg, and other physicists stuck to their respective methodological guns even when it became clear that their characteristic ways of doing physics had both positive and negative features. The timing turned out to be crucial in terms of the significance of the product of the individual approaches. Trying to "flatten" these respective approaches and regard the flattening as refined analysis is unwarranted; it runs the risk of concealing both the methodological dynamics and the key differences. In general, a refined scientific practice means perfecting a specialized tool for investigation and dividing the labor among those using it. This is equally true of the group of physicists who created quantum theory and the group who created quantum mechanics. Both groups practiced each of these approaches to an extent—for example, Heisenberg used experimental results as a starting point for devising his mathematical model, and Bohr produced formulas in the end—but, as we will see, every physicist preferred one approach to the others, especially when faced with crucial dilemmas and problems. Yet the interplay of these different approaches led to the formation of quantum theory and, later on, quantum mechanics.

Bohr's gradually crafted experiment-oriented method was not the only bottom-up approach to physical phenomena;[12] other distinct approaches of that sort were practiced by Sommerfeld, in one stage of his career, and by a host of other experimentally minded laboratory leaders like Thomson and Rutherford. But Bohr's variant of that approach turned out to be more ambitious, fully realizing its capabilities as it aimed at very general hypothesis as an outcome of a gradual climb from experimental hypotheses via intermediary ones. And, perhaps most surprisingly for the reader familiar with the philosophical scholarship on Bohr, this method was at the heart of his distinction between classical and quantum concepts.[13] The distinction was first and foremost an important implication of a division between experimental and theoretical stages in the inductive process and his understanding of their relationship, not a result of an obscure metaphysical commitment or a philosophical idea perhaps learned at an early stage, as is often argued.

Thus, if successful, master hypotheses such as Bohr's are bound to be at odds with deeply entrenched metaphysical and intuitive biases on which theoretical expectations are predicated. In fact, mainly because of the independence of the experimental hypotheses from which the higher-level ones are generated or to which they are attuned, the general account in the form of a master hypothesis (Bohr's atom, or his complementarity principle) is rarely if ever fully in accord with the models of microphysical phenomena that stem from ready-made metaphysical principles. These are typically either rejected or substantially transformed. This was exactly the case when the basic concepts of old quantum theory were emerging, from 1900 to the mid-1920s, and further on, when quantum mechanics was being developed. Thus, Bohr's construction of the model of the atom as a master hypothesis—the most general hypothesis available at the time—his development of the correspondence principle as a constructive and versatile intermediary hypothesis, and, a decade and a half later, his crafting of the complementarity principle as a new master hypothesis were all carefully elicited from a comprehensive grasp of the available experimental context. They did not result from outlandish metaphysical and epistemological views. In fact, this sort of level-headed inductive process deliberately and systematically avoided metaphysically motivated or mathematically driven hasty generalizations based on partial experimental evidence. Part 3 of this book will make clear that, unlike Bohr, Schrödinger and to some extent Heisenberg pursued such generalizations to the potential detriment of the ongoing development of the theory. It was Bohr who tamed these rather hasty generalizations and synthesized them into his general hypothesis. It is thus not so much that during the bottom-up phase of the quantum revolution, physicists "deftly shifted between different pictures of the reality as it suited the tasks at hand" (Sebens 2020, 42). Rather, being firmly entrenched in the experimental context, physicists either created new pictures by modifying old ones (synthesizing new ones from bits and pieces of the old), tried to disregard the old ones entirely, or finally tried to stick to them at any cost.

*

Finally, there are numerous lively debates on induction in philosophy of science (Henderson 2018); and even after centuries of discussion, the jury is still out on whether there is a uniform inductive method of generating scientific knowledge. These debates include a wide variety of views on what constitutes the inductive process, ranging from those basing it on the material facts to those emphasizing the sufficiency of very abstract inference rules and principles. And it is not only the views of professional philosophers on the notion of induction that are relevant, but also those of some prominent scientists like Einstein or Sommerfeld who reflected on this issue. Discussing the details of these numerous diverging views is not within the scope of this book. What is important for our purpose, is that, on the one hand, conceptual discussions sometimes veer off from the complexities of actual science, and even when they don't, they are typically tested against one or a very few somewhat detailed examples. On the other hand, a detailed case study can hardly warrant a general claim about the inductive method. Yet its results can agree with a particular general account or family of accounts, or serve as a counterexample to them. And if a case is central in the history of the development of a scientific field, this adds weight to the analysis and its arguments.

I should note an emerging trend of disparaging case studies in the field of the history and philosophy of science because of their supposed irrelevance to general philosophical concerns relating to science, as well as their alleged methodological impotence, based as they are on a handful of examples. Philosophically motivated analysis of episodes in the history of science emerged as a subfield of philosophy fairly recently and is still in its infancy. It is too early to make any grand assessments. Yet thoroughly and adequately researched cases have to be an anchor of any general philosophical accounts of certain features of the scientific method, and of science in general. This is true for the philosophical understanding of the inductive process. A properly laid out case analysis, based on the facts and their comprehensive and coherent understanding, should be informative to those seeking a general account of induction. Moreover, case studies may be the only way to grasp inductive processes taking place in revolutionary times; our generalizations and ready-made philosophical views seem very likely to fall short.

In any case, detailed historical studies are an invaluable resource for developing various aspects of general philosophical accounts of the way scientific knowledge is produced. Ideally, case studies and abstract general arguments with views they support should come together. Integrated historical and philosophical analysis should be a meeting point, generating a clear account in a historically limited but detailed context. This sort of convergence and coordination of historical and philosophical analysis could minimize hasty conclusions about something as complex as the scientific method. We don't have to stick to the old credo that history of science without philosophy of science is blind, or to the credo that integrating history and philosophy is a "marriage of convenience" to the detriment of both, enabling neither good philosophy nor good history of science.[14]

It is a truism that these two fields cannot exist without each other. However, a particularly fruitful convergence and eventual symbiosis of the two involves an interpretation of how particular scientific fields or theories developed. As Jutta Schickore (2011, 456) states, "Philosophical reflection on science is interpretive, and . . . historicist analysis of scientific, epistemological, and methodological concepts augment our understanding of science." Analysis starts with a provisional concept (e.g., hypothesis, induction, theory, experiment), and through subsequent interpretation, "both the provisional concept and the historical record are elucidated and clarified" (ibid., 457).[15] It is fair to say that I have aimed at a "procedure through which preliminary concepts and points of view and initial judgements are brought together and modified and adjusted until a cogent account is obtained" (Schickore 2011, 472), much in the spirit of Catherine Z. Elgin's (1996) description of the interpretive process as achieving "reflective equilibrium."[16]

In line with this general approach to the interpretive integration of history and philosophy of science, this book is neither an elaborate conceptual discussion with an extended case study as an illustration, nor an extended summary of the current conventional wisdom about the history of early quantum mechanics. This is a historically informed and philosophically motivated interpretation, not a full-blown historical account of Bohr's work and his interactions with his peers. I have discussed the details of some of its key historical aspects in various papers (Perović 2017, 2013, 2008, 2006, 2005), and will point the reader to them when it helps achieve the main focus of this book: a comprehensive view of Bohr's methodology. Thus, this book represents an integrated historical and philosophical perspective on the work of Niels Bohr, in which particular philosophical doctrines are put in parentheses, so to speak, and historical analysis is allowed to lead the way in an exploratory manner. As a result, the book offers a detailed outline of the structure of Bohr's approach to physics; a bottom-up process of generating hypotheses about microphysical (quantum) phenomena.

Taking historians' work seriously in philosophical analysis can tie the analysis to the practice of science, rather than to certain philosophical preferences. In the present case, such an approach precludes the all too common reliance on Bohr's speculative mature thought, removed as it is from the experimental basis on which his main contributions were developed. Moreover, considering "classical concepts" in Bohr's work—a key question in its interpretation—in what follows, I return to the historical accounts of key episodes in the early stages of quantum theory when the notion of "classical physics" initially emerged. As I see it, this offers a plausible starting point to understanding the notion as it was embedded in Bohr's practice of physics, and in his reflections on that practice.

I should also mention that my reluctance to use a particular inductive model stems from my strong impression—gained from my study of the history of quantum mechanics, as well as particle physics—of the existence of several quite different phases in the development of the theory, each characterized by the prominence of a substantially different approach to physical phenomena, especially in terms of the relationship between experimentation and theory. Each of the existing inductive models seems to adequately grasp only one of those phases; any overarching account of induction spelling out the relationship between facts and theory should take this into account. It is partly for this reason that I have not committed myself to any specific view of induction. Rather, I only think of it here in very general terms, as a matter of the central role that experiments play in generating hypotheses *at all levels* of theory formation. This approach to microphysical phenomena is firmly centered in Bohr's vision of physics. We do not need to ask whether there is a unique inductive method, or to identify it in order to explore the inductive/experimentalist nature of Bohr's approach to physics. And, as criteria in the debates on induction have certainly evolved over time, we need to see how exactly they are rooted in the past (Schickore 2011, 460). Generally speaking, I think the philosophical study of the scientific method is more akin to biology than to physics, in the sense that for each claim we make, we can be pretty sure that an exception can be found if the net is cast widely and deeply enough (in the past). Yet general tendencies, pathways, and developmental phases can be identified within specific periods or specific episodes.[17] This is exactly what I aim to do.

Despite these cautionary notes and disclaimers, a question is likely to be asked by those dwelling on the exact nature of induction in science: How did hypotheses come about in this concrete case? Was the inferential process led by facts, or by a set of rules of inference? What we can say is that Bohr favored the bottom-up approach: facts and experimental results—or, more precisely, the particular experimental hypotheses that resulted from careful deliberation on the experimental context—led the way, while other theoretical, formal, and metaphysical aspects were secondary. To show this, I go on to explore and explain the methodological role of experiments in Bohr's work within the overall development of quantum theory and quantum mechanics up to the 1930s, showing how they gradually shaped the generation and testing of hypotheses of different scope. In chapter 10, I also summarize Bohr's experiment-driven approach in the form of a few explicit guidelines, heuristic rules of sorts, for building a web of mutually supportive lower "local" (experimental), intermediate, and general hypotheses within an array of experiments. In other words, I offer an analysis of the structure of the process that generated Bohr's main results, which turns out to be inductive-hypothetical (especially in contrast to the approaches of some of his prominent peers), by interpreting his practice and his reflections on it. How exactly this connects to the existing philosophical debates on induction and the existing views figuring in them is another important matter, one I cannot discuss here.[18]

The developments discussed in this book occurred a century or more ago. Despite the centrality of quantum mechanics in contemporary cutting-edge theoretical and experimental fundamental physics, the issues surrounding these developments are increasingly seen as antiquated in the study of the history and philosophy of physics. They are giving way to numerous exciting developments in contemporary physics. Yet not only can they ground current attempts to understand inductive process in science and the scientific method in general, but the form of some philosophically charged debates in physics, like the one on quantum gravity, is driven by similar underlying methodological issues and tensions (Esfeld 2019). It is far too soon to dismiss Bohr and his contemporaries as irrelevant.

PART 1

Preliminaries

2: FROM LABORATORY TO THEORY

Bohr states with great precision the kind of interplay between theory and experiment so typical for the days of the old quantum theory.

—Abraham Pais (1991, 192)

Bohr's curiosity about philosophical issues started in his youth. As a student at the University of Copenhagen, he witnessed and participated in frequent philosophical discussions with Harald Høffding, a notable Kantian philosopher and a family friend who frequented the Bohr residence, and took a compulsory course in philosophy with him. During his undergraduate years, Bohr was also a very active member of a lively circle, Ekliptika, a group of enthusiastic students who regularly gathered to discuss philosophy. The circle discussed epistemological, metaphysical, and other philosophical issues, and Niels and his younger brother Harald played a prominent role in it. Bohr exhibited a conversational style of argument, sparring with his brother on philosophical matters and on the predicament of reaching a communal understanding (Rozental 1968, 25–26). Harald became a renowned mathematician and Bohr's intellectual collaborator (ibid., 16).

Bohr's early interest in philosophy was matched by his passionate interest in experimentation and lab work. This fact is typically ignored in philosophically driven discussions of early influences on Bohr's philosophical vision of physics. His attitude toward science was influenced as much by his experimental practice as by his philosophical interests, and he continued developing those interests throughout his life. This chapter will thus provide a brief orientation, setting Bohr's career in the context of major trends in the balance of theory and experiment in his discipline, and explaining principal milestones in the development of his work.

The early influence on Bohr of his father's close friend, the physicist Christian Christiansen, was decisive. Christiansen was a notable international physicist and a leading Danish physicist in several areas, but he most successfully studied electrocapillary phenomena. In fact, he was a staple in the Bohr family house—as was Høffding (Pais 1991, 99).

Bohr's formative practical experience in experimental science started in his father's home laboratory during his undergraduate studies. His father, Christian Bohr, was a very successful and quite famous physiologist who worked on experiments tackling physical processes that enable physiological activities. Niels Bohr had an advantage over other students in physics at the University of Copenhagen; the experimental facilities there were meager at the time, but he could work in his father's laboratory (Rozental 1968, 32). In fact, his first notable scientific work was an experiment he performed in this laboratory. His study of the surface tension of water by observing a regularly vibrating jet demonstrated his outstanding abilities as an experimentalist, won him a gold medal from the Royal Danish Academy of Sciences and Letters, and resulted in his first publication (Bohr 1909) after he sent a modified version of the manuscript to the Royal Society in London. Since there was no proper laboratory at the university for Bohr's experiment, Christiansen lobbied for one to be built but was unsuccessful. Bohr then constructed the apparatus in his father's laboratory by blowing glass into small tubes that produced elliptical jets. It was an exceptionally done experiment, with an innovative apparatus and measurement solution: "The experimental part of his gold-medal research, performed in his father's laboratory, required dexterity in glass blowing and photography, the ability to design and assemble a complicated apparatus, and the elaboration of protocols for exacting measurements" (Aaserud and Heilbron 2013, 154). A commentator at the time offered praise: "This work proclaims its originator's special pleasure and ability at working theoretically on problems" (Rozental 1968, 32).

After he obtained an undergraduate degree in physics, Bohr continued his career in Thomson's laboratory and later in Rutherford's. Laboratories such as theirs were centers of the development of physics and made major contributions to the emergence of atomic and quantum theory. The work in the labs was defined by a continuously flourishing synergy of theoretical and experimental work, of which Thomson and Rutherford were essentially the overseers and coordinators. Bohr was especially impressed by Rutherford and his style of running a laboratory, but both men had exceptional talent for recognizing promising experimental and theoretical approaches and phenomena to be investigated. This kept the theoretical and experimental work moving.

Thomson's Cavendish Laboratory at the University of Cambridge was one of the two most prominent centers in physics; the other was the Physico-Technical Institute in Berlin. Thomson's lab gathered together some of the key physicists who were working on the electron theory, including James Jeans and Joseph Larmor. It is quite possible that Bohr decided to spend time in Thomson's lab over labs in continental Europe because "Thomson was more alluring . . . , prolific in ideas, clever in mathematics, and playful in physics" (Heilbron 2013, 47). Under his direction, the laboratory was a lively leading center.

In his time at the lab, Bohr was involved in experiments with cathode rays but soon retreated to his own studies. He found the laboratory quite confusing, and his experiment was not successful (ibid., 25–26). It was a rewarding stay, however, as he became acquainted with certain experimental techniques and the ideas of some prominent physicists, especially Samuel B. McLaren, one of only a handful of physicists who argued that the classical framework of physics could not account for radiation phenomena.

The real flourishing of Bohr and his immersion in the community of experimentalists coincided with his move to Rutherford's laboratory in Manchester. A number of physicists who became instrumental in the development of quantum theory and atomic theory were working there on cutting-edge approaches, most notably on Rutherford's model of the atom. Two physicists who made a contribution to Bohr's work on atomic states in Manchester were Charles Galton Darwin and George de Hevesy. The latter was a renowned experimentalist and became his close friend, while the work of the former, a grandchild of a much more famous grandfather, on experiments with X-rays and alpha rays was a decisive early influence.

Even though he did not share Bohr's optimism on the emerging model of the atom, Rutherford made a great impression on Bohr as a laboratory leader. Bohr once said Rutherford was "almost like a second father" (Pais 1991, 129). Judging by the descriptions of some of his closest collaborators, such as Heisenberg, his characterization of Rutherford is almost identical to his own leadership style, a role he assumed soon after his Manchester experience: "When I turned to Rutherford to learn his reactions to such ideas [on atomic theory], he expressed, as always, alert interest in any promising simplicity but warned with characteristic caution against overstating the bearing of the atomic model and extrapolating from comparatively meager empirical evidence" (Bohr 1961, 1083). The last part of this sentence, warning against overgeneralizing to a model of a physical phenomenon on the basis of inconclusive evidence, outlines one of the key methodological principles Bohr assimilated into his role as mediator between experimental work and the construction of abstract theoretical models. This ability was combined with Bohr's recognizable style of formulating a problem by laying out existing points of view reflecting seemingly disparate accounts of studied physical phenomena as clearly as possible in order to assess the possibility of their synthesis, which was already apparent in an early series of lectures on radioactive transformations (Rozental 1968, 35).

*

The symbiosis of experiment and theory characteristic of Rutherford's, Thomson's, and then Bohr's own laboratories was typical at the time. Physics was largely understood to owe its success to its status as an experimental science. In that sense, throughout his career, Bohr was a physicist firmly situated in a laboratory, acting as a mediator between theoretical and experimental advances that fed off each other continuously, practically daily.

The philosophical and conceptual confusion caused by some of Bohr's work may be a result of the misunderstanding of how he saw his role as a physicist. Some harsh critiques stem from anachronistic expectations of what a leading physicist was supposed to produce. The figure of a theoretician who rarely enters the lab but leads by successfully theorizing—later embodied in Richard Feynman, Julian Schwinger, Steven Weinberg, and others—was not dominant. It may be somewhat of an exaggeration to state that "at the beginning of [the twentieth century] the separation between purely experimental and purely theoretical engagement had just barely begun" (Pais 1991, 101), given the way James C. Maxwell, Ludwig Boltzmann, or Josiah W. Gibbs practiced physics, but the concept of a theoretical physicist who develops impressive coherent mathematical models while being far removed from experimental work was firmly institutionalized only in the second half of the twentieth century. Measured by the standard applied to this way of practicing physics, Bohr's work may seem inferior and even outright strange. Yet his role was that of a typical laboratory mediator—a leading figure in physics at the turn of the twentieth century—which quickly turned into a wider role as a mediator at the level of the physics community working on quantum phenomena. When we understand that he excelled in this sort of role, only then can we properly understand how his work was generated, what its goals were, and what role it played at the time.

Léon Rosenfeld, an early biographer of Bohr's, also a major figure in modern physics and Bohr's collaborator, refused to be drawn into views of Bohr's physics as premised on erroneous and anachronistic assumptions, and described the nature of Bohr's work more appropriately with respect to the practice of physics at the time:

Bohr did not draw any sharp distinction between theoretical and experimental research; on the contrary, he visualized the two aspects of research conducted in such a way as to give each other support and inspiration, and wanted the outfit of the laboratory to be such as to make it possible to test new theoretical developments or conjectures by appropriate experiments. In order to keep up with the changing outlook of the theory, it was imperative, in this conception, to expand and even to renew the experimental equipment in order to adapt it to entirely new lines of research; this Bohr did with remarkable foresight as well as persuasive tenacity in securing the necessary funds. It was a part of his activity to which he devoted much and attached much importance, and the tradition he thus founded continues to bear fruit today (Rosenfeld in Bohr 1972–2008, vol.1, xxx–xxxi).

In short, Bohr followed a long tradition of laboratory physics, and excelled in his role. As I will demonstrate, his theories emerged from the bottom up, through a painstaking and slow inductive process, starting with careful considerations of data in the context of actual measurements and the apparatus on which they were generated, and ending with experimentally all-encompassing provisional accounts of phenomena.

In fact, the experimentalists developed theories of their own about the phenomena they and others studied experimentally; they did not necessarily rely on theorists for help. This is particularly obvious when we assess the way the experimental context affected debates on microphysical phenomena—for example, the debate on the wave and particle features of quantum phenomena where the scattering experiments played a major role. I will assess this episode in more detail, but it suffices to note at this point that J. J. Thomson, Arthur H. Compton, and Darwin, whom we may see today as prototypes of experimentalists, developed detailed theoretical accounts of microphysical entities and their interactions, as did Bohr (along with Hans Kramers and John C. Slater) on the one hand and Einstein on the other. Similarly, Rutherford's model of the atom and his crucial insight that positive charge was concentrated as a nucleus in the atom were direct results of his experiments; in fact, the insight eluded physicists who were focused on devising general theoretical accounts.

Typically, the experimental work was directly motivated by the pet theories experimentalists developed and was related to the theoretical accounts of others more indirectly. This is why, for instance, it took some time for Bohr and Schrödinger to sort out exactly how novel experimental results with the scattering of light by electrons affected their respective accounts of microphysical phenomena (Schrödinger's wave-mechanical account and the Bohr-Kramers-Slater theory) in the mid-1920s. At the time, the boundary between theoretical and experimental work was not drawn sharply across the community, and there was little sense of expert specialties.

In terms of what he practiced in the laboratory and in a wider physics community, as well as in terms of his own reflections on the scientific method, Bohr was a physicist firmly entrenched in the bottom-up experimentalist strand of the scientific method. That strand also marked the first phase of the development of quantum theory and quantum mechanics, up to the mid-1920s. Other concerns, such as metaphysical limits or mathematical elaborations of the physical theory, the focus of some other prominent physicists at the time, came a distant second on Bohr's list of priorities and on the lists of other laboratory leaders such as Thomson or Rutherford.

Throughout his career, Bohr continually reflected on the ongoing experimentally-led pursuit that resulted in the old quantum theory, and then in quantum mechanics. His visions of physics and of its actual practice in the community that developed quantum theory and quantum mechanics were inherently related. His vision was interwoven with the predominantly experimentally-driven development of quantum theory and quantum mechanics until the mid-1920s. For him, as for Sommerfeld, who made critical contributions to Bohr's model of the atom, "experiment was a constitutive element at multiple stages in the production of theoretical work" (Seth 2010, 3). Moreover, as was customary in this strand of pursuing physics, much like Bohr, "not merely deploying experimental data as a final point of comparison for theoretical results, Sommerfeld made use of this data during the process of constructing his mathematical expressions" (ibid., 150). This contrasted with Planck's rather conservative approach of composing a well-rounded theoretical schema (modeling statistical properties of the system with the help of the data) and at times a romantic idea of common principles he thought physicists had pursued in the past. More specifically, although Planck crafted his quantum theory of radiation in close connection with the experimental data (Kangro 1976), he was not as keen on using the experimental results as a key tool as were Bohr and Sommerfeld, for the most part. It is not surprising, then, that Planck introduced into his theory of radiation infinitely small resonators that were not meant to be tested experimentally (ibid., 173).

This is why Bohr's work, including his main contributions, is best approached in a wider context. Methodologically speaking, I am following Don Howard (1994) to a great extent in my analysis, in order to explain the emergence of Bohr as an experimentally minded lab chief and mediator of the quantum physics community. Bohr's vision was inspired by the approach to physical phenomena taken by Thomson, Rutherford, and other experimentalists whose work was crucial for the development of the theory in the first two decades of the twentieth century. It stands in stark contrast to Schrödinger's metaphysically oriented approach, and to that of a younger generation of physicists, including Heisenberg, Dirac, and Pauli—the generation who took the baton of developing quantum physics from Bohr in the 1930s. This latter generation was primarily concerned with mathematical formulations of microphysical phenomena—a vision far different from that advocated by Bohr. These emerging theoreticians were not competent in the experimental laboratory work; nor were their models built in close connection with the experimental work. Reportedly, Heisenberg failed to competently respond to any questions about experimental tasks or experimental equipment at a job interview. Bragging about such incompetence even became fashionable among theoreticians at the time—a practice introduced famously as the Pauli effect, after Pauli, who established the bragging rights (Seth 2010, 182). The way they crafted their hypotheses, their theoretical structures, and their goals all reflected this.

*

Bohr's first major contribution to the debate on the nature of microphysical states was the model of the atom he announced in his 1913 publications (Bohr 1913a, 1913c, 1913d). The model famously combines the idea, motivated by the classical-mechanical planetary model, of electrons moving along the orbits around the nucleus, with the discontinuous nature of their energy states as they are assigned to only certain orbits and transitions between these orbital states in the form of quantum leaps that explain radiation and absorption processes. The idea conflicted with the main presuppositions of classical mechanics— most prominently, the continuity of energy in microphysical processes. Bohr's second major contribution during the same period, the principle of correspondence, an analogy relating classical-mechanically treated frequencies of electrons to frequencies of transitions between higher orbits (with higher energy values) in the atom, was used as a limited but supportive intermediary step in inducing the model of the atom.

A variety of novel experiments in the mid-1920s, as well as the emergence of wave-mechanical and matrix-mechanical formalisms, led the community to expect a refurbishing of the existing quantum theory that centered on Bohr's two contributions. As a stream in this development,

Bohr gradually introduced his concept of the complementarity of classical and quantum states, together with the notion of wave-particle duality. The main goal of the complementarity account, his third major contribution, was a pragmatic reconciliation of wave and particle approaches to quantum phenomena, both partial but inescapable in accounting for particular features of experimental phenomena. It was a heuristic move that aimed at devising a new provisional central hypothesis from the updated and rather unexpected experimental results.

3: FROM CLASSICAL EXPERIMENTS TO QUANTUM THEORY

In this chapter I explain the experimental context in which Bohr worked, and give an account of his gradually devised conceptual approach to experimental apparatus and research in physics at the most basic level. Two distinct stages in Bohr's vision of physics provided the foundation for his main contributions and were also apparent in his reflections on the practice of physics and his gradual development as a physicist: the stage of inducing the basic lower-level hypotheses from the experiments, and the stage of inducing higher-level theoretical hypotheses via intermediary hypotheses. Understanding the exact structure and role of and relationship between these stages is the key to understanding Bohr's work and his vision of physics in the overall context of emerging quantum mechanics in the first three decades of the twentieth century. Such an understanding will also prevent us from falling for any exaggerated claims of the influence of particular philosophical ideas on Bohr's work.

Generally speaking, this division has been an essential part of the modern scientific method. At some point, any process in science aimed at generating adequate hypotheses about physical phenomena has to involve the gathering of experimental, mostly numerically expressed data and the relevant features of the experiments that produced them. In this respect, physics at the turn of the twentieth century was a standard scientific enterprise pursued in small laboratories where theory and experiment were in constant flux, substantially different from large experiments a few decades later, when exceedingly demanding experiments started lagging behind theory. Bohr's approach to physics belongs in this general methodological context; but more specifically, and unlike that of some other prominent physicists, it reflects the bottom-up construction of hypotheses. Yet he crafted his approach, his vision of how physics should be pursued, during the rise of quantum mechanics. Consequently, his practice of physics and his understanding of it—in particular, his grasp of the relationship between theory and experiment—are inseparable from his distinction between the classical and quantum world. The distinction was an important implication of a way of grasping the division between experimental and theoretical stages in the specific context of studying quantum phenomena. It was not a result of an obscure metaphysical commitment.

This distinction has been a major stumbling block in studies of Bohr's work and has triggered a plethora of criticism, with some identifying his view as the pursuit of "irrationalism" (Joos et al. 2003; Joos 2006), others characterizing it as "idiosyncratic remarks about the role of measuring devices and the boundaries of theoretical domains" (Bitbol 2017, 48), and still others calling it inadequate in the light of quantum decoherence accounts (Bacciagaluppi 2012; Schlosshauer 2004; Zurek 1981; Zeh 1970). Although the comment on decoherence may be valid, there is a more fundamental and prosaic methodological way to approach Bohr. To grasp his understanding of the scientific method, to appreciate his three groundbreaking achievements, and to assess their exact role within the overall dialogue in the physics community, we certainly need to grasp his understanding of the role of classical physical concepts and their relationship with the quantum concept. Now, this aspect of his work, just like the nature of his main contributions, becomes noncontentious if we keep in mind the overall inductive process to which it is tied.

Thus, first, the classical physical notions are inherently tied to Bohr's understanding of the first stage of the inductive process—the process of observing, recording, and reporting experimental particulars in the lab—while quantum concepts are unavoidably inherent, in Bohr's view, to the second stage. Elsewhere (Perovic 2013, 166–67), I develop the point about the methodological nature of Bohr's distinction and his insistence on classical properties in the context of the experimental process as distinct from the further, higher-level theoretical inferences. And more recently, Dieks (2017, 310–11), Kristian Camilleri (2017), and Camilleri and Maximilian Schlosshauer (2015) have made a similar point.[1]

Second, it is important to emphasize that Bohr's qualifications of measurements and experimental apparata in the second, theoretical phase do not necessarily stick to everyday accounts amended with the basic language of classical physics. It is not unconditionally true that "no acceptable account of what is observed in a laboratory, and no acceptable description of the instruments that are used for that purpose, can be given in nonclassical terms according to Bohr" (Bitbol 2017, 50). This statement applies—for quite obvious reasons, as we will see—to the first, experimental stage of the inductive process which elicits lower-level hypotheses removed from high theory and theoretical models, but not to the second phase.[2]

In the remainder of this chapter I will provide the basic framework for understanding the classical/quantum distinction as an implication of

Bohr's understanding of the scientific method as he approached it, by explicating and further clarifying his views. In later chapters, a fuller account of this common thread across different phases in Bohr's career and work will be offered in a historically sensitive analysis of his major achievements. But before moving on with the discussion of the main points just outlined, I should note that, though Bohr had thought about and discussed with his peers the epistemic ramifications of experimentation and theory in physics all along, his explicit and more elaborate views appeared in the published work only gradually. As we will see, he formulated these views as he worked out the physics problems he was occupied with. This is, in fact, something we might expect, given his priorities as a physicist.

Until the late 1920s, many of Bohr's explicit reflections on the practice of physics, his own and that of the entire community, were present in the form of occasional albeit insightful formulations in his papers and books aimed at the physics community and primarily at quantum physicists. In fact, the analysis of his approach to and understanding of physics while he was building his model of the atom and up to the development of the complementarity must focus on the published work he targeted at the quantum physics specialists simply because his longer substantial reflections lack this kind of understanding. And this should be pursued along with the analysis of his actual practice, his correspondence, and various accounts of his peers. Longer substantial, explicit published discussions appeared more regularly only in the late 1920s and early 1930s. In the 1930s he started targeting audiences beyond the quantum specialist community with various publications. For example, his essay "Life and Light" is written for a broad scientific audience, and is quite far removed from his actual practice of developing various aspects of quantum physics, in sharp contrast to, for example, his Como lecture or the 1928 *Nature* piece that preceded it, which was written for the physics community keen on understanding emerging quantum mechanics. In the 1940s and 1950s, when Bohr was already past his prime as a physicist, so to speak, and a few decades after his major contributions, he dedicated substantially more time to general topics of philosophical and cultural significance. Understandably, such work was less connected to his experience of lab physics. Thus, we should bear in mind that the clarity and astuteness of Bohr's arguments, and what we can and should draw from them, will vary accordingly. All these various works are a great resource for understanding his vision of physics during the quantum revolution—how he thought it worked exactly, and the basic elements of its method—but we should distill an account from these varied resources carefully, especially bearing in mind the character of the argument and the target audience, and in concert with the analysis of the actual practice of physics in all of the major phases of his work when he was making his major contributions. This chapter gives such an account, and squares it explicitly with the actual practice of physics in the rest of the book.

In general, a very cautious synthesis of the resources is needed if we wish to produce a full picture of Bohr's vision of physics. One possible way of skewing the analysis (see chapter 15) is to treat Bohr's later general philosophical arguments targeting a general audience as a guide to his entire oeuvre, when his early specialist papers and the historical context can actually tell us more about how he understood and practiced physics. Similarly, Bohr started developing his notion of classical physics and classical physical states, discussed in the remainder of this chapter, when he was devising his model of the atom in 1913. This is reflected in his early works aimed at quantum physicists as much as in his explicit general statements on that subject in the 1920s and 1930s and later, including his publications for a wider audience; my forthcoming analysis takes this into account.

*

Inferences at the first stage of the inductive process in basement-room type laboratories happen at the level of the experimental setup and the selection of observations and measurements deemed suitable—that is, at the level of gathering *experimental particulars* via the experimentalists' senses and the immediate reports on them. In the first comprehensively articulated account of the inductive process in science, Francis Bacon conjointly labels the experimental particulars the "effects" (Bacon 2000) and "the works" (Bacon 1874)—the components of the experimental machinery, its mechanisms and things the experimenter observes, and measurements in them. Allan Franklin (1989) and Peter L. Galison (1997; 1987) have developed thorough accounts of the role played by the structure of the experimental apparatus in the pursuit of specific epistemological goals in physics, and the ways it affects the pursuit of scientific knowledge.[3] Yet the main point Bohr makes about the nature of the experimental apparatus, the induction of the basic level of knowledge from it, and the subsequent use of this knowledge to build the theory is rather general. The following passage summarizes this view of the ground level (i.e., the experimental level) of the scientific process:

> By the word "experiment" we refer to a situation where we can tell others what we have done and what we have learned [when experimenting] and that, therefore, the account of the experimental arrangement and of the

results of the observations must be expressed in unambiguous language with suitable application of the terminology of classical physics (Bohr 1949, 209).

Thus, Bohr first insisted on explicitly characterizing the nature of experimental particulars ("experimental arrangement" and "the results of the observations") gathered and recorded during the first stage in as precise a manner as possible. The nature of our perception is such that we are confined to an observational context and everyday language that expresses it, at the ground level of physical knowledge, amended with the terminology of classical physics. In "On Atoms and Human Knowledge," published in 1958, he states:

> All unambiguous information concerning atomic objects is derived from permanent marks—such as a spot on a photographic plate, caused by the impact of an electron—left on the bodies which define the experimental conditions. In the analysis of single atomic particles, this is made possible by irreversible amplification effects—such a spot on a photographic plate left by the impact of an electron, or an electric discharge created in a counter device—and the observations concern only where and when the particle is registered on the plate or its energy on arrival with the counter (Bohr 1958b, 169).[4]

Second, Bohr had emphasized the use of suitably *amending* classical descriptions in such endeavor: "The aim of every physical experiment leaves no choice but to use everyday concepts, perhaps refined by the terminology of classical physics" (Bohr 1939, 269), as they describe the apparatus, its manipulation, and the "actual experimental results." At the height of the creation of quantum mechanics and his complementarity account, which replaced his model of the atom, Bohr said that any scientific theory, including quantum theory, begins with the help of classical concepts employed to account for experimental particulars. Thus, "the unambiguous interpretation of any measurement must be essentially framed in terms of classical physical theories, and we must say that *in this sense* the language of Newton and Maxwell will remain the language of physicists for all time" (Bohr 1931, 692; emphasis added). The adamant tone, more typically absent from Bohr's usually reticent pronouncements, stemmed from the fact that this and similar statements clarified what Bohr considered the ground level, the experimental level, at which the foundation of any physical theory is constructed.

It is the level of the experimental records and reports in experiments that must be sorted out. Reports on experimental particulars should be expressed in common language, amended with common classical technical terms when needed. The reports aim at a reduced and unambiguous content, and formulations and language have to match this aim. The accounts that pick out particular, simple observations and experimental arrangements in the lab must be composed of simple unambiguous reports of what the experimenter's senses selected and recorded.

The statement quoted above and others similar to it often cause interpreters to go off on a tangent, assuming that Bohr is talking about the full-blown interpretation of quantum states and is reflecting some kind of higher-order metaphysical statement. But this last comment of Bohr I have quoted, published in 1931, is interpreted quite naturally when read in conjunction with Bohr's previously quoted comments on the nature of experimental apparatus and observations. He uses the same wording and addresses the same issue. The quotation from his 1931 paper perhaps best summarizes his view of the role of experiments. There, Bohr explicitly limits his claim about classical descriptions to the experimental context using the phrase "in this sense."

We really need surprisingly little interpretive work to understand Bohr's viewpoint when it is placed in the proper context: that of gathering and describing experimental particulars. This is the case in part because the notions of classical physics and classical physical states, which Bohr suggests should simply amend unambiguous everyday descriptions of experimental particulars, have been customary for over a century now. At the time, however, these notions were novel and only just emerging. The notions of "classical mechanics" and "classical thermodynamics" as terms subsuming particular physical theories and theoretical concepts were initially introduced around the turn of the twentieth century, and their meanings were extended to include electrodynamics, statistics, and physics only in the first two decades of the century (Staley 2008). These notions reflected a number of key debates in physics at the time.

Now, the milestone meaning of the notion of classical physics that was assimilated in quantum theory was probably crafted but certainly cemented at the Solvay Council in 1911 (ibid., 308–9). By that time it was clear that the principle of the equipartition of energy, whereby there is an equal partition of energy across oscillating particles that radiate, had to be abandoned in light of quantized oscillations introduced by Planck. At the conference this was touted by Planck himself as an unambiguous indication that "classical mechanics, fructified and extended by electrodynamics" (Planck 1913, 77), was not up to the task of accounting for the nature of microphysical states and processes. And a number of experimental results that could not be reconciled with existing theory became subsumed under the label "classical physics" (Heilbron 2013, 27; Aaserud and Heilbron 2013, 144).

Bohr was only indirectly informed of the results of the 1911 Solvay Council. He was a young physicist at the time, and the notion of classical physics defined in contrast to emerging quantum theory made a great impression on him, as we will see in the following chapters. But he was well acquainted with and had thought about debates on classical physical states, and amalgamated them in his breakthrough work. As a result, his notion of classical physics and classical physical states was very different from Boltzmann's treatment of classical states as an amalgamation of the commitment to atomism and statistical mechanics, the purported foundation of the basic understanding of physical states as such. As we will see, Boltzmann's view greatly influenced Schrödinger's interpretation of quantum mechanics in the mid-1920s, contributing to contention between Schrödinger and Bohr.

Bohr aimed at a notion of "classical" that captured the basic, experimental level of doing physics, thus putting quantum states not prone to equipartition, or to accounting in strictly discrete terms, in their proper place: outside the domain of the experimental work generating initial results (experimental hypotheses), and at the level of hypotheses derived from those results. Unlike Boltzmann's programmatic notion, Bohr's notion of the classical physical states and classical physics had a distinct pragmatic aim of clarifying the methodological framework of quantum physics. The notion of the classical physical states Bohr gradually built and explicated is crucially tied to the experimental practice—in particular, to the notion of everyday terms used to describe observations in the experiments. As such, it reflects how his overall vision of physics informed his research program in quantum physics, and how his practice as a physicist shaped his overall vision. Here, his epistemological reflection and the practice of physics clearly came together to form a practice-based yet epistemically refined approach to physics.

Furthermore, what are commonly called the "experimental results" to which Bohr also refers actually include everything that is selected and recorded—"experimental arrangements" and the process ("what we have done"), not simply the bare numerical data—that is, everything later assimilated into more general hypotheses on the phenomenon at stake. In the experimental physics of the first half of the twentieth century, this

level of the scientific process is clearly reflected in journal reports on experiments, with detailed descriptions of experimental equipment and its
use: the common aspects of discussions among physicists such as Bohr,
Rutherford, and Thomson.

Subtler derivations later in the process are stating the underlying conditions of producing the numerical results. Bohr was aware of this, and he
invoked the nature of the apparatus when making more refined theoretical
points. The information about the place and time of discharge created in
a counter device, for example, "presupposes knowledge of the position
of the photographic plate relative to other parts of the experimental arrangement, such as regarding diaphragm and shutters defining space-time
coordination or electrified and magnetized bodies which determine the
external force fields acting on the particle and permit energy measurements" (Bohr 1958b, 169–70).

Overall, "the functioning of the measuring instruments must be described within the framework of classical physical ideas" (ibid.), the
framework involving unambiguous discreetness and localization of physical states and principles like equiparition of energy. Thus, what we may
characterize as the overall *experimental context* that grounded the hypotheses was not simply a stream of numerical values produced in various
experiments, but included various aspects of the experimental setups and
processes that produced these values, while probing various physical phenomena presumed to be inherently related. This is crucial, but, given the
empiricist individual-centered epistemology of science, it is often overlooked in understanding the role of experiments in inducing hypotheses
and accepting or rejecting theories. The simplified picture of evidence *e*
simply confirming hypothesis *h*, or theory *T*—an abstraction perhaps useful in some philosophical contexts—is too abstract to help us understand
the nature of the inductive process in experimentation and, accordingly,
the key divisions of labor in the physics community (Boyd 2018). This is
especially true of the early-twentieth-century physics in which experimentation and theoretical work were closely intertwined.

Thus, at the first stage of the process of inquiry in physics, the observations of experimental particulars are gathered and expressed in "common language" (Bohr 1948, 313). "Common language," or the collection
of everyday concepts that describe the regular physical world around us,
is in fact a starting point for further refinement of concepts in classical
physics. Yet everyday language is often not sufficiently precise in its characterization of the properties of the gathered experimental particulars.
So an experimental particular—or, in Bohr's words, "the experimental

arrangement," or "the record of the observations of experimental situations" (ibid.)—will be put into the helpful shape of "a well-defined meaning in the sense of classical mechanics" (Bohr and Rosenfeld 1933, 359).
The language of classical mechanics has distilled precise physical concepts from the everyday language descriptions that experimenters can use
when recording experimental particulars.

*

What exactly is gathered in these experimental reports, and in what shape
is it? Spectroscopy played a key role in the assembly of Bohr's model of the
atom. In their accounts of these experiments, the experimentalists noted
spectral lines on the screen, including their intensity and the order of
their distribution. In experiments with cathode glass tubes, also essential
to Bohr's model, the experimentalists reported the distances between the
plates and glass, as well as the glow itself, its intensity, its hue, and its
location, along with the level of purity of the vacuum and the quality of
vacuum pumps that produced it (figure 1). As an amendment to the use of
regular notions such as glow, distance, intensity, and location, the notions

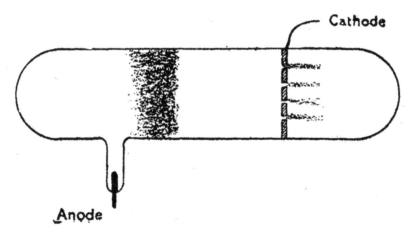

Figure 1. J. J. Thomson's 1913 drawing of light beams streaming through the holes
of a perforated, charged cathode in a vacuum tube. The color of the light streams
behind the cathode depends on the gas applied in the tube. This phenomenon
("Kanalstrahlen") was not well understood before Thomson's experiments and
the development of quantum theory.
From J. J. Thomson, "Bakerian Lecture: Rays of Positive Electricity," *Proceedings of
the Royal Society of London. Series A, Containing Papers of a Mathematical and Physical Character* 89, no. 607 (1913): 1–20. Permission conveyed through Copyright
Clearance Center, Inc. Republished with permission of the Royal Society (UK).

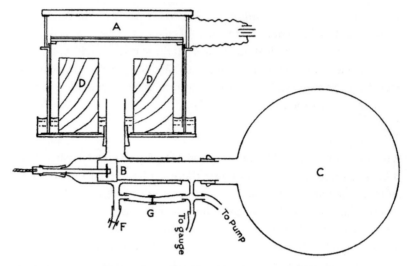

Figure 2. C. T. R. Wilson's 1912 drawing of his invention of the cloud chamber, which went on to define experimentation in particle physics for the next several decades. The tracks of charged particles in the cylindrical chamber (A) filled with the saturated vapor are photographed and analyzed to determine the properties of various particles. The vacuum tube (C), connected with the chamber via a valve (B) and a regulating wooden cylinder (D), controls the extent of vapor saturation. Two clamps (G and F) control the incoming air volume. Wilson initially photographed the vapor ionization tracks (water droplets condensing on ions) produced by the incoming X-rays and alpha rays.
From C. T. R. Wilson, "On an Expansion Apparatus for Making Visible the Tracks of Ionising Particles in Gases and Some Results Obtained by Its Use," *Proceedings of the Royal Society of London. Series A, Containing Papers of a Mathematical and Physical Character* 87, no. 595 (1912): 277–92. Permission conveyed through Copyright Clearance Center, Inc. Republished with permission of the Royal Society (UK).

of charge and discharge were customarily used to describe the operation of such experiments and their aspects deemed significant. A few generations earlier, those notions had not been part of the operating language; but they were newly introduced through a series of experiments with electricity, as part of the general hypotheses accounting for a score of experimental results. Similarly, the tracks were recorded and measured in Charles T. R. Wilson's cloud chamber (figure 2) and their quality noted (whether or not they were continuous; whether or not they were split and how; whether they were straight or curved and at which angles; figure 3). The notions of pressure and temperature were customary classical physical amendments

Figure 3. One of the first photographs to show the tracks of subatomic particles in a cloud chamber. The picture was taken in 1911 by the cloud chamber's inventor, the English physicist C. T. R. Wilson, at the Cavendish Laboratory in Cambridge. The tracks are caused by alpha particles emitted by a small amount of radium at the top of a metal tongue inserted into the cloud chamber.
C. T. R. Wilson and Science Photo Laboratory.

to the everyday language in the basic accounts of such experiments. And in the double-slit experiments, the dots on the screen were observed; their positions recorded; the energy of the source determined; and the characteristics of the light ray, the screen, and the double slit described.

Bohr labels all these items "experimental accounts" of "experimental arrangements" and "results of observations." But prior to examining this notion in detail, it is important to note that here I use the notion of the hypothesis, or of postulations broadly understood, as a general working term to cover the entire inductive process, from these basic accounts of experimental particulars observed and recorded in laboratories (lower hypotheses) to theoretical models of phenomena, the notion of theoretical principles or axioms, and finally the notion of theory as a comprehensive and substantially mathematized structure grasping relevant phenomena.[5]Bohr and his contemporaries used each of these terms in various contexts and for various purposes, including for notions of the "model of the atom,"[6] the "correspondence principle," and "quantum theory." They sometimes used them interchangeably, especially for the notions of principles and axioms. Yet all these concepts are hypothetical postulations, and their exact hypothetical nature depends on their proximity to experimental particulars. Accordingly, we label them as lower, intermediate, and master hypotheses.

But perhaps most importantly, even the accounts nearest the experimental particulars that Bohr singles out as "accounts of experiments" are hypotheses, albeit simple ones. The way Bohr formulated his characterization makes this apparent. As accounts of the experiments, they tell us "what we have done and what we have learned" in the experiments; they surpass the "naked" observations of the experimental process to give an "unambiguous *interpretation* of . . . measurement" (emphasis added). These accounts "of the experimental arrangements and of the results of observations" are unambiguous; they are formulated in everyday language and amended with notions of classical physics where appropriate. But they are interpretations, too—selective accounts (i.e. already interpretive, albeit minimally, if compared to more general hypotheses of the second theoretical stage) of experimental "arrangements." They involve selecting and relating certain observed experimental particulars, as well as selecting particular (everyday and classical-mechanical) notions to account for them.

The measurements and experimental particulars are embedded in such lower-level hypotheses, articulated in everyday language and augmented with the notions of classical physics when necessary. An experimental account, then, is a hypothetical synthesis based on selected experimental particulars. For instance, Heinrich Hertz's (1883) account of experiments with cathode rays in the gas tube reported the lack of their deflection by the magnet (selected particular experimental observational aspects), while Thomson singled out this particular aspect of the experiment *along with* the (higher) level of gas exhaustion in the tube.[7] A lower hypothesis in light of other experimental results—that is, other lower hypotheses—will typically turn out to be deficient or streamlined by certain initial preferences as the inductive process unfolds, while higher-level hypotheses will aim to reconcile all these lower hypotheses. Thus, Thomson's account of the experiment involving a particular selection of the experimental particulars it related was really an experimental hypothesis that turned out to be richer and more precise than Hertz's. Thomson tested it further, and found out that the lower levels of gas in the tube enabled deflection of the rays, thus rendering Hertz's formulation of the hypothesis rough and deficient. The case demonstrates the slow and gradual ascent from Hertz's hypothesis to Thomson's, from one experimental hypothesis based on particular recordings of the experimental situation and its multiple replications to another, along with all the intermediary steps like Perrin's (1895) experiments and their treatment by Thomson.

The spectroscopic experimental work that was crucial to the induction of Bohr's model of the atom (especially the rules established by Johannes Rydberg and Johann J. Balmer) was also the source of experimental data for the development of quantum theory as a whole. Later, quantum mechanics convincingly demonstrates the hypothetical nature of lower-level experimental accounts and inferences during their gradual and selective inclusion into the emerging higher-level hypotheses. The physicists working on spectroscopy at the time differed to the point of convoluted controversy on which exact patterns should be selected as significant and why. Their views differed both on which lower-level hypothesis grouped lines most adequately, and which intermediary hypothesis provided the best concepts for capturing them. Each competing representation made a particular selection of recorded experimental particulars—that is, the splitting of spectral lines and the conditions under which they were produced, grouped in a particular way (Kragh 1985; Carazza and Robotti 2002; Seth 2010, ch. 7). For instance, the so-called fine structure of the regular spectral lines produced by hydrogen resulted in a long-standing controversy. A number of experiments indicated that the seemingly regular line in the spectrum was, in fact, separated. Yet "the actual appearance of the doublet"—that is, whether the line was characterized by the separation or not—"naturally depends on the *relative intensities* of the individual lines" (Kragh 1985, 71; emphasis added), —that is, how exactly the experimenter

Hβ
1A. = 43 mm.

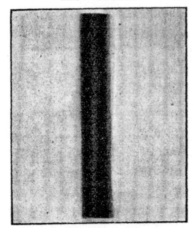

Ha
1A = 34·5 mm.

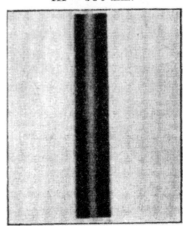

Figure 4. T. R. Merton's photographs of doublets, proposed at the time to be a feature of the fine structure of the Balmer series of hydrogen spectral lines. This hypothesis turned out to be correct after a heated, decade-long debate and multiple experiments. Merton attempted to stabilize the "doubling" under varying conditions to determine whether it was a stable feature of the lines or the result of "impurities." He labeled the behavior of the lines "very capricious."
From T. R. Merton, "On the Structure of the Balmer Series of Hydrogen Lines," *Proceedings of the Royal Society of London. Series A, Containing Papers of a Mathematical and Physical Character* 97, no. 685 (1920) 307–20. Permission conveyed through Copyright Clearance Center, Inc. Republished with permission of the Royal Society (UK).

will judge relative intensities of the lines, and whether she will deem them an indication of the existence of the separation, or an artifact of some sort (figure 4). This was the point of controversy for almost a decade, and various lower-level hypotheses concerning the separation of the line were debated; different experimenters accounted differently for the phenomenon, suggesting opposed hypotheses composed of differently selected observational elements and experimental conditions.

In fact, there were three series of such small revolutions in experiments with spectral lines. The first concerned the lines and their distribution (the work of Rydberg and Balmer was crucial here), the second doublets, and the third the multiplets in their fine structure. It would be hard, perhaps even impossible, to understand spectroscopy in this period, the course of the debate, and how the results were assimilated into more general hypotheses crucial for the emergence of quantum theory without understanding their hypothetical nature: that they subsumed selected observational and other experimental aspects (or facts, if a reader is keen on using that notion, albeit in a very precise sense, as embedded in lower hypotheses) in the form of fairly simple experimental hypotheses that could be improved on or deemed skewed or irrelevant in due course.

Thus, when studying the inductive process that led to the quantum theory, especially Bohr's contributions to it, we need to take into account that relevant pieces of experimental knowledge which are frequently and customarily labeled in other philosophical, historical, or scientific contexts as "experimental laws" and "facts" coming out of the laboratories—or, to use Bohr's more cautiously chosen term, "accounts of experiments"—had a hypothetical character at the time they were assimilated into the theory.[8] And the lower-level hypotheses were being assimilated into higher-level hypotheses all along. Physicists were not waiting for the experimentalist community to complete its work, as it were, in order to build or test their theories. Often, the final word on whether the work was completed, and which experimental results were adequate (and could perhaps be pronounced as relevant experimental facts) came at the height of the theory, certainly with mixed results. And the inclusion of the lower-level hypotheses into the theoretical structure continued even as more experiments were forthcoming. Thus, as we will see, Philipp Lenard's (1903) model of the atom did not require more precise results than those from Hertz's experiments, while Bohr's model took advantage of Thomson's more refined hypothesis. And the physicists working on the atomic structure at the time viewed the rules accounting for the distribution of spectral lines proposed by Rydberg (1890) and Balmer (1885) as little more than numerological rules for counting observed spectral lines (Carazza and Robotti 2002, Bohr in Rozen 1968, 51), a rather initial phase of an experimental work. They did not bother to try to relate them to their theories of the atom and radiation, but Bohr used them to refine his own model of the atom and made them a crucial piece of evidence for it.

*

All in all, in the experimental process stemming from Bohr's characterization, the lowest experimental hypotheses are drawn from fairly simple but selective observations of experimental particulars; the hypotheses are generated from selected observational "facts" or "arrangements" (the works), as Bohr labels them. They gather the data, along with particular features of the experimental apparatus and the accompanying processes during the data production. In terms of generating adequate hypotheses about the experimentally probed phenomena, all this constitutes the primary stage.

Once we understand that Bohr insisted on the experience and observed phenomena expressed in unambiguous language as the groundwork for the gradual induction of hypotheses, it should no longer puzzle us that he deemed this stage of crafting a theory to be isolated from quantum concepts; the development of the latter, including its concepts and its language, firmly belongs to the second inductive stage. The first stage takes place in relative separation from the second stage, the stage at which physicists leave their labs, as it were, to gradually devise more substantial multiple hypotheses from multiple lower hypotheses (i.e., experimental data and relevant experimental features generated in the labs). The theorizing capabilities of the scientists can fully kick in only once they set out to connect the lower hypotheses and bring them under more general and substantial ones.[9]

Bohr's insistence on everyday language *cum* classical terminology accounts of experiments in the first stage of induction looks perfectly natural. Quantum descriptions can enter accounts of the experimental apparatus as an afterthought, but not during the basic collecting and sorting out of experimental particulars in the lowest hypotheses during the first stage. This is a matter of the limits of our senses and the language that captures the experimental particulars we gather with the senses. As the notion of "classical states" is characterized here, it contrasts with the characterization of quantum states, devised at the second stage. Thus, the first stage of the inductive process is inevitably restricted to unambiguous reports of measurements: intensity and distribution of spectral lines, intensity and location of the glow in a cathode tube, dots on the screen, tracks in the cloud chamber, clicks of a Geiger-Muller counter and recordings of their separation in time, acknowledgements of simultaneous events or events separated in time, and additional more complicated but basically similar experimental particulars. The concepts of momentum, position, force, or field potential in Newtonian mechanics and electrodynamics stem from such common language describing measurements. The discreteness and localization of momentum and position are invariably there; they are the basis of our everyday observations and our reports on them. Once we compare their work to quantum theory, Isaac Newton and Maxwell seem to have been fortunate that properties introduced in their theoretical concepts did not have to depart from the properties of localizability and discreteness of observations and immediate reports on them.

As we will see in the following chapters, Bohr's commitment to and the exact view of the distinction gradually evolved as he developed his contributions to "old" quantum theory and then to quantum mechanics. Yet in fact, Joseph Larmor, who worked at Cambridge during Bohr's tenure there, and whose book *Aether and Matter* (Larmor 1900) Bohr revered, was already discussing Maxwell's struggles to couch the theory of electromagnetism in new terms, noting that it was a process that looked haphazard and confusing and was essentially unfinished. He insisted that this was an inevitable stage of the development of theories, especially theories of the microphysical world that must borrow the notions of ordinary mechanics.[10] Bohr must have learned this lesson at Cambridge, and could hardly have expected that a quantum theory of the microphysical world would fare any better, at least in the phase of laying down its foundations, guided by numerous experiments.

With the advent of quantum mechanics, it became clear that quantum states cannot be straightforwardly formulated in terms of individual states that simply assume discrete values; a microphysical state with momentum p lacks a discrete value of position, and vice versa. The account of such a quantum state was understood to involve both the state of the observer and the state of the object. Similarly, the state of an entangled quantum system depends on spatially and causally separated states. In contrast, the position, momentum and force in Newtonian mechanics and Maxwellian electrodynamics are unambiguous in the sense in which our everyday observations are unambiguous; they cannot be quantized or entangled, nor can they mutually exclude one another. Nor can the experimental particulars. Both experimental particulars and classical states are, to use contemporary terminology, localized and separable (or discrete),[11] and that is why we can borrow classical physical terms to characterize experimental particulars in a more precise fashion. In an attempt "to avoid misunderstandings [Bohr] preferred not to talk of an observation as influencing the object of investigation" (Klein 1968, 92); this was a confounding of Heisenberg's work in the 1920s and the so called-Copenhagen interpretation of quantum mechanics, as I will discuss later. Leaving aside theoretical inferences for the second stage, Bohr opted "to use the word 'phenomenon' to describe observations gained under specific conditions including an account of the whole experimental arrangement" (ibid.).

Bohr makes a clear distinction between the first and the second stage of induction, where only the latter can involve quantum principles and inferences. In the following passage on the observer-object entanglement motivated by Heisenberg's indeterminacy principle, Bohr sets the experimental particulars on one side and the level of the hypothesized elicitation of such entanglement on the other:

> We must on the one hand, realize that the aim of every physical experiment leaves us no choice but to use everyday concepts. . . . On the other hand, it is equally important to understand that just this circumstance *implies* that *no result* of an experiment concerning a phenomenon that lies outside the range of classical physics *can be interpreted as* giving information about independent properties of the objects (Bohr 1939, 269; emphasis added).[12]

Put otherwise, the aim of higher-level hypotheses is to tell us how these classically accounted-for experimental particulars captured in terms of the lower-level hypotheses truly relate to each other. We might subsequently characterize the apparatus itself in quantum terms, as Bohr himself occasionally did when debating various theoretical points of view, for the purposes of assessing various higher-level general hypotheses. But these characterizations inescapably belong to the second inductive stage, not to the building blocks of lower hypotheses assembled from experimental particulars.

The second inductive stage exceeds the stage of gathering and sorting out experimental particulars and can involve the concept of the entanglement of observed objects and instruments, quantum entanglements, or nonlocality. The first stage consists of "the application of these [classical] concepts alone"—the concepts entirely isolated by the very nature of the process of gathering experimental particulars from the concepts introduced at the second stage—to the gathering of particulars which then "makes it possible to relate the symbolism of the quantum theory to *the data of experience*" (Bohr 1934, 16; emphasis added). Klein similarly describes Bohr's account of the first stage: "Unambiguous conclusions from observation require that the experiments themselves be describable by means of classical physics, i.e. that effects connected with Planck's quantum of action may be neglected, as far as the measuring arrangements are concerned" (Klein 1968, 90). Only then can we take a second, theoretically-minded look at the performed experiments, the nature of the apparatus, the observer, and other similar aspects while considering the use of various concepts (including quantum concepts) for such a purpose. This look may contain attempts to relate the measurements, expressed in classical language, to a range of tentatively suggested intermediary and general hypotheses that may eventually produce a comprehensive grasp of the overall experimental context, a master hypothesis. Such hypotheses can introduce metaphysical ideas and models that, in the end, do not agree with the classical nature of basic states. In fact, physicists eventually had to embrace the quantum hypotheses that violate the requirements of separability and localization of physical states and introduce the entanglement of the observed objects and instruments in interpretations of them.

Again, this was not a new challenge. For instance, Larmor (1900, vi) said the criticism of his ether theory of the molecular and microphysical world had to "leave reality behind"—that is, the reality directly accessible to the senses. His argument was essentially that, once the formulation of the higher–level hypotheses begins, such challenges are ill-advised; the novel notions in the hypotheses can be certainly challenged as a flight from reality, but "so in fact may every result of thought be described which is more than a record or comparison of sensations" (Larmor 1900, vi)—that is, the bottom level. The second stage of the inductive process has its own rules, so to speak, and the scientist needs to understand the nature of its independence from the stage of data gathering to develop adequate theories through induction.

The second stage requires the construction of new concepts which depart from direct records of experimental particulars and thus can exceed the limits of classical concepts of lower hypotheses. Before the emergence of quantum theory, the higher-level hypotheses in physics unified the lower hypotheses, its everyday and classical notions, by devising new classical concepts. But during the emergence of quantum theory, the entire inductive structure became unusually apparent, especially the distinction between the two stages, as limitations of language tied to experimental observations became obvious, and hypotheses of the second stage had to introduce nonclassical quantum concepts to account for discrepancies between lower-level hypotheses. Bohr insisted that quantum phenomena brought to light this unavoidable limitation of the language of lower hypotheses in a particularly explicit manner, repeatedly noting (as Heisenberg paraphrases him), "How satisfying it was that this limitation had already been expressed in the foundation of atomic theory in a mathematically lucid way" (Heisenberg 1968, 107).

Bohr consistently denied that "the fundamental concepts of the classical theories will ever become superfluous for the description of physical experience" (Bohr 1934) at that first inductive stage of experimenting. Insofar as the basic experimental level was concerned, he said, the account of experimental particulars, or "the experimental arrangement and the record of the observations of experimental situations . . . must always be expressed in common language supplemented with the terminology of classical physics" (Bohr 1948, 313). Bohr never changed his mind on this core aspect of the inductive process; he wrote these two comments over the span of almost fifteen years.

Even though Bohr's viewpoint on what he thought were classical properties that characterized experimental particulars and recorded experience is clear, we might still question whether and to what extent classical physics—electrodynamics in particular, despite it being characterized as "classical" early on by Henri Poincaré and Larmor (Staley 2008, 302)—has relied on ideas that accord with the discreteness that characterizes experimental particulars. For example, is a wave front discrete, even though it is localizable?

Once we account for the wave front in classical electrodynamics by treating its components as localized and discrete, we may understand the entire wave front as a localized and discrete compound. Nothing suggests that this analogy cannot be pursued in individual processes that underlie the dynamics of the wave front. As a matter of fact, Bohr was clear: "The activity of electromagnetic field quantities rests by definition on the transfer of momentum to suitable electric or magnetic test bodies" (Bohr and Rosenfeld 1933, 368). In this case, such a body is treated as an average element of the electric field, spatiotemporally localized, assuming discrete values, with no additional limitations on its activity.

These analogies appear unsatisfactory in subsequent hypothesis building in the case of quantum phenomena, including the nature of the experimentally examined "individual processes" in the mid-1920s that I will discuss later. And even though we are able to, for example, experience a hazy light incidence in experiments, we have a tendency to report on such occurrences as individual events that may only afterwards be connected in multiple ways to a single event observed at the same time but observationally disentangled from it. The haze may be examined further, but the reports on the state of its elements will be provided in discrete terms (e.g., intensity, color of the components). Arguably, we do not always observe discrete values—cognitive scientists are the ones who will ultimately tell us whether we do—but we can attempt to pick out such states in our observations and record them as such. For instance, we may observe the dissipation of light, but we will analyze the process in discrete terms.

Finally, and most importantly, we can certainly describe and explain the experimental apparatus in quantum terms to clarify various theoretical points. Bohr himself occasionally used this method in debates and in writing (Camilleri and Schlossauer 2017). He did not confine his theoretical understanding of the experimental apparatus to classical physics. Yet, crucially, this sort of description belongs to the second stage of induction and, as such, is induced from and thus firmly grounded by classically formulated lowest hypotheses that account for experimental particulars.

Galison (1997) introduced a distinction between experimental instruments using logical circuits—that is, binary modes—and those using analog (continual) detectors. His analysis grasps mainly experimental instruments after Bohr's initial breakthroughs and the formation of quantum mechanics. Bohr's strict commitment to classically formulated experimental results and classical observations, which in effect meant discrete experimental values, may have been the result of a very specific experimental context at the time, and perhaps it has to be loosened in light of subsequent experimental work in physics. Moreover, the increased automation of the experimental process in modern physics, especially after World War II, including detection, recording, and analysis, renders this particular distinction irrelevant as the experimental results can be captured in various formal frameworks.

Although I will not consider it here in detail, it is an important question whether and how exactly quantum concepts entered lower experimental hypotheses and experimental reports after the 1930s. Was it in a form that violated Bohr's bottom-line view that all reports and accounts must be classical? In fact, the first possible indication that lower-level hypotheses may have been perceived as adequately accounted for in the language containing the notion of quanta was the work of Sommerfeld on deriving theoretical inferences from the data of spectral analysis in the 1920s. The semi-empirical kind of inference led Sommerfeld to state that "the regularities that here obtain throughout are primarily empirical in nature, but their integral character demands from the outset that they be clothed in the language of quanta" (Sommerfeld 1923, v). It is not clear however, whether this characterization is simply an attempt, a trick of sorts, to bypass the intermediary hypotheses (primarily the correspondence principle) that bridge lower-level hypotheses with a master model, in line with the doctrine of nonmodeling direct phenomenological inferences which Sommerfeld pursued at the time. We need to look past the 1930s to find more systematic uses of quantum descriptions in characterizations of the lower-level hypotheses.

In particular, accounts of quantum decoherence (Zurek 1981; Zeh 1970) were motivated to a great extent by criticism of the insistence on the inevitability of the concept of classical states. Yet an apparent inadequacy or irrelevance of the concept in characterizing experimental situations as mere instances of decoherence taking place generally in quantum states may not be a methodological point at the ground level of the experimental process Bohr had in mind. And even if we can use quantum concepts at that level, it is hard to see how the quantum theory and quantum mechanics could have been built *ab novo* without the gradual process that included Bohr's distinction to start with. An emerging theory and a mature theory should not be judged by the same standards.

*

We should bear in mind that in his writings and discussions on the method of physics, Bohr was reflecting on his practice in the lab and in the physics community. His theoretical reflections and his practice had been in flux since he designed and performed his first experiment in his father's garage. This is largely why his methodological and philosophical accounts present us with a very specific style of reflection: essentially a running commentary, a reflection on his own work, and the work of the entire quantum physics community throughout his career. It is unavoidably a style of thought and written reflection different from the academic, polished, well-rounded accounts of scientific method and induction we find in contemporary work or in the writings of nineteenth-century philosophers. Moreover, Bohr continuously revised his work; Heisenberg called it "a continuous process of improvement, change and discussions with others" (Heisenberg in Kuhn 1963). This was a tortuous continual process, as Bohr himself characterized it (Kragh 2012, 193).

We can thus certainly identify certain insightful overlaps with the points put forward by the authors advocating, for instance, abduction, starting with Charles Sanders Pierce—or, much more recently, by those developing the account of inference to the best explanation. Whatever the overlaps, however, the fundamental level of Bohr's account is his unambiguous insistence on the double-stage nature of the inductive process; this is the baseline of the inductive process.[13] Finer points on that process, as well as reflections on metaphysical presuppositions and mathematical articulation of hypotheses, are bound by this fundamental level of analysis: it's a starting point and an anchor of Bohr's thought.

A few other authors (Bitbol 2017; Dorato 2017; Chevalley 1994; Kaiser 1992; Hooker 1972) have noted this sharp two-stage distinction as foundational to Bohr's approach to the physical world, though they disagree on its origin and its underlying philosophical grounds. Despite their differences and their attempts to read it through particular philosophical views (mostly Kantian) rather than seeing it as a regular aspect of the inductive-experimentalist method in science, their assessments and conclusions put Bohr's work in a much more plausible context than do many other attempts to interpret its nature and aims.

In fact, we can fully understand the double-stage nature of induction and its importance once we look at the wider context that gave rise to Bohr's contributions, and at the arguments Bohr pursued in debates with Schrödinger, Heisenberg, and Einstein, all of whom were focused on metaphysical presuppositions or mathematical formulations, and whose work presented a stark contrast to Bohr's inductivist concerns. Bohr's methodology ignores the Machian appeal for economic summaries of experience, or the building of mathematical tools from it (as practiced by Heisenberg and a number of physicists of the younger generation), the appeal also in accord with Sommerfeld's early engineering approach to physical phenomena and his later direct phenomenological reading of hypotheses from experimental results. It is also at odds with Planck's insistence on testing well-rounded theoretical frameworks, Planck and Einstein's insistence on clear and easy-to-grasp general physical principles that ought to govern the study of physical phenomena, and Schrödinger's commitment to particular intuitive metaphysical principles.

Without keeping in mind that Bohr's arguments are grounded in and substantiated by a distinct sort of understanding of the ascending inductive process in physics, and that other aspects of his thought are its further refinements, his arguments may look obscure or weak, or may even appear to us as a form of intellectual bullying, as they have to a substantial number of authors—that is, if we are tempted to seek what is not there and was not supposed to be there. Admittedly, although the fundamental points of Bohr's view are apparent in his work, that work is not a well-rounded account. In what follows, I will fill in the gaps by explaining the context of the development of quantum theory while showing how Bohr's reflections on this context shaped his vision of physics.

PART 2

Bohr's Vision in Practice: The Old Quantum Theory

4: SPECTRAL LINES, QUANTUM STATES, AND A MASTER MODEL OF THE ATOM

The question is not only of the development of the interpretation of experimental facts, but just [as] much by means of these [facts] to develop our deficient theoretical conceptions.

—Niels Bohr (1972–2008, vol. 3, 397)

This type of analysis that was partly based on an amazing skill in separating effects according to orders of magnitude was characteristic of all his work. In this respect he was much closer to experimental physics than more formal theoreticians.

—H. B. G. Casimir (1968)[1]

Bohr insisted that any hypothesis put forward at the stage of developing a theory, be it a full-scale theory, a model, or a principle, should be developed in light of, or ideally shaped by, experimental results via hypotheses induced from the experimental results, or the lower hypotheses. In an account of his correspondence principle, Bohr says: "The question is not only of the development of the interpretation of experimental facts, but just [as] much by means of these [facts] to develop our deficient theoretical conceptions" (Bohr 1972–2008, vol. 3, 397).

This emergence of new concepts in higher-level hypotheses (understood in a broad sense, as specified earlier) via experimental results happens gradually. As we have seen, the hierarchy of hypotheses starts with those nearest the experimental process that pick out the experimental particulars directly. Next come the intermediary ones, with a broader but still limited grasp, semidirectly connected to the experimental context. Finally, the master hypotheses aim at a broad adequate grasp of the entire experimental context: as wide a scope of relevant experimental results as possible across all the relevant experiments. In a step-by-step fashion, the hypotheses raise the extent of generality and abstraction and result in what we might label a full-scale physical theory. Master hypotheses aim at encompassing all the relevant phenomena explored in all the relevant experiments by accommodating all the antecedent (supportive) intermediary hypotheses that concern only certain limited aspects of the experimentally probed phenomena. A particular master hypothesis is central until new experiments are performed, and new lower-level hypotheses question it. The process is thus an inductive loop of sorts. This highest stage of induction resulting in the master hypothesis should be guided first and foremost by lower and intermediary hypotheses, which generate the key concepts and language for the formulation of the master hypothesis. The aim is, "by means of these [facts,] to develop our deficient theoretical conceptions."

As we will see in part 2 of this book, this is how Bohr built his model of the atom. He gradually transformed the concepts and language of lower hypotheses across various experiments he recognized as essential; and via intermediary supporting hypotheses (chapter 4) and the correspondence principle as a constructive versatile intermediary hypothesis (chapter 5), he assimilated them into a general model of the atom. The concepts and language of this master hypothesis successfully linked and unified the entire set of relevant lower hypotheses. We will see in chapters 4 and 5 that Bohr's work on the model was only the final stage of the elaborate inductive-hypothetical process he gradually developed and reflected upon, which extended into the community of physicists working on quantum phenomena at the time. Bohr's reception by the community and the role he assumed as a result of his contributions are the focus of the last two chapters (6 and 7) in this part of the book. In fact, he devised his account of complementarity later on (the topic of part 3), as the next step in a continuously evolving inductive process.

*

Until the mid-1920s, the focus in the development of quantum theory was on the performance of multiple experiments and the induction of adequate hypotheses from a growing experimental base. Bohr excelled in this phase of the development of the theory. In 1911, about the time he started crafting his model of the atom, he wrote to a colleague that he was "very enthusiastic about quantum theory," adding parenthetically, "I mean its experimental side)" (Bohr 1972–2008, vol. 1, 431). The second, mathematically driven phase started in the mid-1920s and increasingly required a different focus and different skills. The earlier experimental context had been comprehensive, resulting in a number of seemingly discrepant lower hypotheses that Bohr's model of the atom ultimately synthesized. Bohr's remarks on the two-stage inductive approach to physics, effectively a reflection on the ongoing process, are often reduced to his summaries and descriptions of his struggle at various steps in the emergence of quantum theory from the experimental base. We need to look at the details of this inductive process to appreciate how immensely challenging the task was.

Perhaps the most challenging part of this inductive process was the fact that the observing, selecting, and recording of particular aspects of experimentation in the small laboratories of the late nineteenth and early twentieth century was essentially shielded from deep theorizing that produced aspiring master hypotheses. In these nonautomated labs, the experimenters were inevitably confined to fairly simple and limited acts of measurement and limited observations due first and foremost to practical constraints. The observation of experimental setups and the experimental process in such labs is quite unlike the later-stage induction of intermediary and master hypotheses: in labs where the experimenter is overwhelmed by sensations and various concrete features of the experiment, she has to select and record.[2] Thus, as she begins perceiving and recording, the experimenter already and inevitably starts creating "lower" hypotheses, nearest to the experimental particulars. She focuses on recording various aspects of the experiment, including the details of how the experimental phenomena that manifest themselves to the senses are produced. The senses, however, do not seem to be simply impartial. Rather, in their default state they are quite dull, limited, and deceptive, as Bacon and other early theoreticians of experimentation had already noticed. Although this may be an exaggerated characterization of the nature of our senses, the insights of modern cognitive science suggest, and Joshua Hughes (1998) and Maria Rentetzi (2007) have argued, that it takes effort and specialized training to be sufficiently vigilant in the laboratory setting.

The recorded experimental particulars and data are initially gathered into the lowest hypotheses from these measurements and observations; the intellectual powers that flourish at the level of intermediary and master hypothesis formation have a limited reach. This is what Francis Bacon called "written experience," the limits of which were explicated by Bohr. From the preceding chapter, recall that the experimenter's written record is limited by language confined to everyday concepts and well established concepts of classical physics as their refinement. The written record allows the intellect to construct lower hypotheses much more precisely than if it relied on memory alone, but the use of classical concepts has a descriptive role in, and sets the limits of the formulation of, the lowest hypotheses close to experimental particulars.

As noted in the previous chapter, this step clearly contrasts with the second stage of connecting lower hypotheses and bringing them under hypotheses entirely outside the domain of the senses. The second stage utilizes theoretical capabilities alone without manipulating actual phenomena, and results in theoretical discussions, debates, and often disagreements. Yet independence from various biases of the senses during their use in these laboratories, and in the formulation of lower hypotheses, is never assured, as perhaps a naive empiricist might expect. The experimenter is never fully isolated from her powers to theorize in the pursuit of lower hypotheses within this classical framework, though the interpretive and explanatory function of lower hypotheses, if any, is reduced first to a selection of one experimental particular over another and, second, to the choice of unambiguous everyday or classical-physics notions. These powers continuously interfere, albeit in a limited way when compared to the second stage where they flourish; and they may nudge the experimenter's choices of experimental particulars in a certain direction. In fact, biases and preferences lurk in the background of lab activity even before the second, more abstract, stage of hypothesis formation, and the experimenter has to actively resist this pull from the very beginning, at the very basic level of collecting experimental particulars and data.[3] The experimenter's struggle to safeguard the autonomy of the experimental process starts at the sensory level: the choice of some experimental features over others from the multitude of possibilities is rarely, if ever, completely insulated from biases and preferences. Formulating the basic lower hypotheses as a report inevitably includes certain preferences that the experimenter should resist.

It is hard not to be aware of these aspects of the experimental work in physics, especially for a physicist like Bohr, given his pursuit of ambitious goals directly dependent on experimental setups. In fact, the relationship between the observer, her knowledge, and the external stimuli she is experiencing via her senses assumed a central role in Bohr's understanding of quantum mechanics after the mid-1920s. Yet Bohr was likely introduced to this central aspect of experimental physics early on, in a systematic and scientifically informed manner. Bohr must have been aware of the sensitivity of human cognitive and perceptual states from his early student days, as his father was a prominent experimentalist who studied the underlying physiological conditions of cognitive processes. In addition, one member of the Ekliptika circle who remained close to Bohr throughout both their careers was Edgar Rubin, a well-known experimental psychologist (Heilbron 2013, 21). In his account of the relationship between mental states and stimuli, unlike the nineteenth-century understanding of the primacy of the mental in psychology, he insisted on the indeterminacy of the responses of the cognizer/observer to external stimuli. A central point of the account was that the knowledge of the observer is prone to influence the response to a stimulus, and will thus influence exactly what and how something is experienced in the observational act and reported on. Moreover, discussions and debates on the nature of perception and its interrelatedness with perceiving and understanding the physical world, the nature of physiological biases and perception, and the understanding of physical states were prominent in the late nineteenth-century physics and were taken up by Ernst Mach, Wilhelm Wundt, Hermann von Helmholtz, and others (Staley 2018). All this must have contributed to Bohr's sensitivity to the process of interpreting experimental facts at the very lowest level of hypothesis formation.

Thus, even the preliminary accounts of experimental phenomena (i.e., lower hypotheses) induced from experimental particulars tend, at least initially, to favor specific theoretical and ontological assumptions and concepts. The choice to design a particular experimental setup is already motivated by certain preferences, and the experimental apparatus is always run by a particular operational theory, irrespective of how simple it may be.[4] The key experimentalist skill, then, in inducing lower hypotheses is to recognize and then minimize or bypass these biases by any means available, ranging from perceptiveness to the use of novel techniques, when producing and choosing experimental particulars and connecting them into lower hypotheses. The skill was essential to the work of Thomson, Rutherford, and the field of spectroscopy, all of which led to Bohr's model of the atom.

In the first phase of the development of spectral analysis, until the breakthrough papers of Arthur W. Conway (1907) and Walther Ritz (1908), the basic assumption of the operational theory in the field was that spectral lines are the consequence of oscillating individual atoms, and molecules are their combination. Based on this assumption, experiments performed in 1865 (Plücker and Hittorf 1865) varied the conditions under which different lines were produced by different chemical substances. But further experiments in the next two decades slowly showed discrepancies in the numerical relationships between wavelengths of various lines supposedly produced by molecules (understood as aggregates of vibrating atoms) of chemical elements, when those relationships were based on the assumption of the model of the atom producing spectral lines because of its vibration (Carazza and Robotti 2002).

Adequate numerical relationships were induced by Rydberg (1890) and Balmer (1885) from the experimental results. These two experimenters showed exceptional skill in sidelining the overall assumption, a bias as it were, of the received atomic theory that shaped the way instruments were constructed and the experimental particulars were treated when the experimenters were inducing lower hypotheses (Kragh 2012; Seth 2010; Carazza and Robotti 2002). These very skills were also pivotal in the formation of the key concepts in Bohr's intermediary (the correspondence principle) and master hypotheses (Bohr's model of the atom).

Relevant experimental particulars in the multitude do not always simply present themselves to the experimenters in a series of experiments; Rydberg's induction of the numerical rule for the distribution of spectral lines is a prime example (Rydberg 1890; Carazza and Robotti 2002, 6). In fact, careful and close focus on the conventionally selected experimental aspects often turns out to be a blind alley. The first three groups of chemical elements in the periodic table were commonly recognized to produce three different combinations of spectral lines: "sharp," "diffuse," and "principal" (also labeled "intense"). This was a standard classification in spectroscopy at the time. Yet Rydberg compared the lines *across groups* and discovered another correlation, effectively a new series of lines that he subsequently subsumed into a suitable formula. The formula crucially introduced the wave numbers as a parameter independent of wavelengths, which in turn enabled a more thorough formal grasp of the distribution of resulting spectral lines in the experiments. This experimental particular, unacknowledged as a significant part of the overall experimental context that saw spectral lines as divided into three main groups, was just one of many particulars that other, perhaps less perceptive, experimenters never noticed or considered significant. This was the point at which the experimenter's observational discipline had to be accompanied by vigilance to, and the resulting isolation of, any underlying bias in the manner in which experimental particulars (spectral lines) were typically recorded. A less astute experimenter sticking to the conventional view of the experimental particulars and the process in the lab, in effect guided by bias, would have never recognized that particular as potentially significant.[5]

The lower hypotheses of this sort were created within a particular domain of the experimental context, set up by the biases that the experimentalists tried to challenge.[6] But unlike in the case of inducing intermediary hypotheses, they were not trying to immediately connect experimental particulars to a specific master hypothesis—certainly not explicitly. Typically, the movement from lower hypotheses to intermediary ones was gradual, as was the building of adequate concepts connecting experimental results with general master hypotheses. The lower hypotheses varied in their scope but, generally speaking, they concerned a limited domain of experimental particulars. Yet the favored conceptualization used in a lower hypothesis aimed at grasping certain experimental particulars in an experiment or set of experiments would typically, either wittingly or unwittingly, encourage generalization across various experimental series. This was the case with Balmer's discovery of correlations in the spectral lines of hydrogen, which in fact motivated Rydberg's work five years later. Balmer speculated on the possibility of generalizing his discovery:

> We might ask ourselves if the previous formula is valid only for a single chemical element, hydrogen, and if it does not appear in the spectral lines of other elements, each with its own fundamental number. If this were the case, we could perhaps assume that the formula valid for hydrogen is a special case of a more general formula which, in particular conditions, becomes the formula for hydrogen's lines (Balmer 1885, 549).

Unfortunately, attempts of this sort often failed, as they attempted to generalize without taking into account other elements of the experimental context. A series of experiments with X-rays passing through gases showed that only a small fraction of molecules of a gas are split up by the ray even though they are uniformly exposed to it. Thomson explained this small quantity of ionization by his "needle radiation" hypothesis: an X-ray front could not be uniform, but penetrated the gas as a series of concentrated "needles" (Thomson 1904, 63–65).[7] Einstein adopted this hypothesis and developed it further into a quantum hypothesis of radiation (Whittaker 1953, 92). Yet this hypothesis was applicable only to X-ray experiments with radiation, unlike a more general one that would adequately account for a wider domain of experiments. The experimental apparatus, consisting of a petroleum lamp and gas tank (Thomson 1903, 258), limited the induced hypothesis to probing only one aspect of the interaction between radiation (X-rays) and matter (molecules of the gas). In light interference experiments, the experimental particulars prevented induction of a needle radiation sort of hypothesis, since even radiation corresponding to a single photon showed directed behavior that was radial-wave-like rather than needle-like. Reconciling this experimentally induced discrepancy required a new concept of radiation and a new model of how it was generated. Only Bohr's synthetic model provided this, and that would come much later.

The eventual failures of such applications in other experiments sometimes not only challenged the initially generalized hypothesis but revealed limitations and significant overlooked aspects in the initial experiments.

As I have mentioned, Thomson's other experiments with X-ray tubes and the hypotheses he drew from them pointed out deficiencies of Hertz's experiments with the same phenomenon, as well as deficiencies in Lenard's model.[8]

Balmer's and Rydberg's new hypotheses in spectroscopy were treated as useful and convenient numerological relationships, but as essentially irrelevant to the higher-level theory of the atom for the time being (Carazza and Robotti 2002; Kragh 1985; Kragh 2012). There was no higher-level hypothesis that could successfully suggest the common mechanism behind these values at the time, nor was it clear that they could be incorporated into one in a significant way. Yet these lower hypotheses slowly became a key element of the change of the basic assumption of spectroscopy—what spectral lines represented in experiments—as this key element was built in the operational theory spectroscopists used. Conway (1907) and Ritz (1908) independently suggested two intermediary hypotheses connecting these lower hypotheses with a rough model of the atom as a disturbed system, with disturbances being due to the motion of an atom's basic elements, not to that of an atom as a whole. In other words, an atom produced spectral lines one at a time in a disturbed state when the parts (electrons) produced a train of vibrations. Building the intermediary hypothesis from the bottom up, Ritz's "point of departure was the formal structure of empirical laws which govern spectra" (Carazza and Robotti 2002, 315)— that is, the lower-level hypotheses of Balmer and Rydberg. The intermediary hypothesis was drawn from the experiments, as an attempt "to identify an internal structure of the atom that could justify them" (ibid.) Once these hypotheses were generated, the spectroscopists started treating the frequency of a spectral line as a difference of two "terms," with the "terms" representing two distinct states of the atom (later connected to the motion of electrons), not the vibration of the entire atom. These intermediary hypotheses introduced the new concept of "terms" or states of the atom as physically significant, representing a crucial step toward Bohr's concept of radiation as a result of the transition between atomic states.[9]

These useful and adequate intermediary hypotheses of Ritz and Conway could not be reconciled with a model of the atom that produced spectral lines as it vibrated as a whole. In an attempt to provide an alternative, Jeans (1901) suggested an early qualitative model of the atom composed of negative and positive electrons in equilibrium, the spectra being the result of oscillations of negative electrons. Electrons were arranged as concentric dipoles on spherical surfaces. Yet the model could not account properly for the frequencies and dimensions of atoms (i.e., distances between charges or spheres) since, as Jeans himself realized, a new parameter was required to make such derivations (Carazza and Robotti 2002).

Meanwhile, other physicists attempted to build models of the atom based on different sets of experiments with electric charges. One set of such experiments (Plucker 1858; Plucker and Hittorf 1865) with cathode rays was interpreted in the context of the nineteenth-century understanding of the wave nature of light. The experiments were designed by analogy to light-interference experiments with obstacles to the glow of cathode rays. Another group of experiments (Perrin 1895) led to the hypothesis of negatively charged rays. A discrepancy between the induced hypotheses started to emerge. Hertz's experiments, mentioned earlier (Hertz 1883), deepened it. Hertz used two charged plates (positive and negative) to attempt deflection of produced cathode rays, but deflection was not observed. As the expectation had been that charged particles constituting the rays would have deflected, the widely accepted conclusion was that the rays instead had to be a radiant propagation of disturbances through ether, analogous to the wave propagation of light as it was understood at the time (Harré 1981; Whittaker 1953).

Experiments with cathode rays that easily penetrated matter led Lenard (1903) to conclude that it was the particles which penetrated the matter (e.g., gas in a tube), but that they had to be uniform in terms of charge, since the matter did not stop their penetration (Whittaker 1953, 22). Generalizing from these results, he devised a model of the atom composed of dynamides as massive charge doublets (dipoles) distributed throughout the atom. The mass of the atom depended on the number of dynamides, but the atom's volume was mainly empty since dynamides were small, and small dipoles penetrated the matter easily. This is an example of inducing a master hypothesis by advancing a new synthetic concept from a limited scope within the overall experimental context. Such inductions do not always succeed; this particular one failed to get any further experimental traction, and died out instead of being adopted as a useful intermediary hypothesis connecting a limited experimental domain with another, more comprehensive model, the way Rutherford and Thomson's model did in Bohr's hands soon after Lenard's attempt.

At the same time, Thomson also performed experiments with cathode rays. He induced a hypothesis opposite to the accepted hypothesis of cathode rays as radial disturbances in ether. It was crucially based on two experimental particulars unnoticed by others in the skillful manner reminiscent of Balmer and Rydberg. "On repeating this experiment [by Hertz]," he said, "I at first got the same result, but subsequent experiments showed that the absence of deflexion is due to the conductivity conferred on the rarefied gases by the cathode rays. On measuring this conductivity . . . it was found to decrease very rapidly with exhaustion of the gas" (Thomson 1897, 296). If the exhaustions are very high, not far from the state of vacuum—in other words, the condition in the glass tube produced by superb vacuum pumps—it is possible to detect deflection, and Thomson indeed measured it. Otherwise, as we now know and as Thomson concluded at the time, the rays will ionize gas and short-circuit the charged plates, and this will be exhibited as lack of deflection. Thomson also discovered that the glow produced by deflection could not be seen in the daylight of the laboratory. The light had to be switched off and controlled in a particular way: "To darken the room to see the phosphorescent patch, a needle coated with luminous paint was placed so that by a screw it could be moved up and down the scale. Thus, when the light was admitted the deflexion of the phosphorescent patch could be measured" (Thomson 1897, 308), and, hence, the angle of deflection could be determined.

The experiments pointed to the limitations of previous experiments performed with Hertz's apparatus. Buchwald (1995) argues that Thomson did not replicate Hertz's original experiments but changed the underlying conditions; the reasons for Thomson's new results had to do with the more refined techniques for producing rays in tubes that Thomson employed in his experiments, not with the density of gas. This account was convincingly refuted by Mattingly (2001) and Chen (2007) as a red herring. Even if we grant an unlikely scenario whereby Thomson substantially changed the conditions of Hertz's original experimental arrangement, despite his above-quoted report, he still drew conclusions from the experimental setup on the dependence of the bending of the ray and the gas exhaustion. He chose to focus on and report on a particular aspect of the experimental setup that Hertz did not consider essential, *though it was part of Hertz's experimental setup as well.* Perhaps Thomson was not looking at the exact same observational setup and drew a different conclusion, though that is unlikely. But he decided to vary the correlation between two components common to their experiments, as he noticed that their relationship might be subtler than Hertz's experiment implied.[10] In any case, the newly noticed particulars questioned the induction of the ray-as-light-wave propagation hypothesis, as Thomson's lower hypotheses introduced the notion of sensitivity to gas density to grasp various levels of deflection. It also gradually introduced the notion of charged particles as more adequate than a radial disturbance that would produce no deflection (Buchwald and

Warwick 2001). This result suggested that the mass of the atom considerably exceeded that of the electrons, when taking into account the actual number of the electrons in the atom.[11]

Rydberg's lower hypothesis took a while before it was assimilated into an intermediary hypothesis of Conway and Ritz, and even longer to be meaningfully related to a full-scale model of the atom in Bohr's model. But Thomson's led fairly quickly to a novel master hypothesis. Thomson's general hypothesis, his resulting model of the atom, introduced the concept of ionization as relevant to atomic structure (a concept already used in understanding molecules), the concept of subatomic charged particles, and a particular distribution of them throughout the atom. The results of Thomson's experiments with X-rays were not sufficient, however, for Thomson to establish electrons distributed throughout the positively charged atom as the key aspect of the model (Whittaker 1953). He based this aspect of his model, first, on an old idea that electric currents feature circular motion, but second, on the more precise notion of the distribution of negatively charged particles across the atom. In this he was guided by Alfred M. Mayer's experiments with the distribution in water of sewing needles with magnetized (as south pole) tips with respect to the large north magnetic pole. Analogously, the entire atom acted as a north pole to the distributed electrons in Thomson's model. Much like the needles in the experiments, the small number of electrons was assumed to concentrate at the same distance from a positive charge, but many electrons were distributed all around as various geometrical formations. Mayer's lower hypothesis was connected to the key feature of the model by analogy—the same sort of move in producing an analog as an intermediary hypothesis that Bohr later used to construct his correspondence principle.

Thomson's model was surprising and did not agree with the central intuition of atoms as wholes, as in the basic dipole carriers of charges that Lenard's model tried to accommodate with his dynamides. Thomson noted, "The assumption of a state of matter more finely subdivided than the atoms of an [chemical] element is a somewhat startling one" (Thomson's speech to the Royal Society in April 1897; reprinted in Thomson 1970, 36). Bacon had already noticed that an upwards construction of a master hypothesis from lower ones is bound to startle standard intellectual expectations as it connects seemingly discrepant phenomena and their aspects (Bacon 2000, 38, xxviii). Bohr took the important piece of the experimental context developed by Thomson into account in his doctoral dissertation (Bohr 1972–2008, vol. 1) when thinking about the structure of the atom, but much more startling "assumptions" defining Bohr's model of the atom were about to be induced when these cathode ray experiments and spectroscopic experiments were considered together.

Rutherford and Geiger's experiments with alpha particles (Rutherford and Geiger 1908a, 1908b) led Rutherford to place the positive charge as concentrated in the center of the atom. In a nutshell, an alpha particle eight thousand times the mass of an electron bounces off a cloud of electrons, indicating that the positive charge has to be concentrated, not distributed around the atom. Rutherford devised this crucial element of his model of the atom by extending the lower hypothesis he had induced from a series of experiments with alpha particles he had performed in 1907. The properties of the so-called alpha rays were puzzling at the time. Rutherford sent them through gas exposed to an electromagnetic field so strong that only allowed the ions produced by the ionization of gas molecules by alpha particles to pass through one by one. These ions were accelerated and magnified by collisions through the electric field, and finally detected—practically on a ion-by-ion basis—by the particle counter constructed by Rutherford's new assistant, Hans Geiger. Given the number and the mass of ions in comparison to those of incoming alpha particles, the results showed that alpha particles must carry a charge opposite to that of the electrons, and must be double in quantity. Later experiments in the series demonstrated that the particles entering the gas were really molecules of helium: the alpha particles emitted by radium interacting with hydrogen molecules showed an immense increase in scintillations, whereas those running through oxygen did not. The spectral analysis of the end product in the interaction determined that it was molecules of helium that entered the gas. From these experimental results and from the results with the deflection of alpha particles at various angles—some particles reached the detector at a considerable angle, and some even turned back—Rutherford inferred the existence of a concentrated atomic nucleus with a mass close to that of the entire atom, and with a charge opposite to that of the electron. The nucleus was tightly packed, rather than spread out in the atom; otherwise, the particles could not be repelled by it. Given the regularities with which negative and positive charges interacted, electrons then had to circulate around the nucleus, forming shells.

Such orbits, however, are not stable if calculated using classical mechanics, as Bohr later discovered, as electrons will lose energy when energy is emitted via the electromagnetic field, since circular motion is an accelerated motion by definition. In other words, an electron circling around such a nucleus would quickly lose its energy, according to Coulomb's law. But the atom is a stable structure; atoms that constitute the same element do not differ in terms of stability under the same conditions. This discrepancy was something that neither Thomson's nor Rutherford's model could explain. Moreover, an adequate model had to explain the finite number of chemical elements, as well as slight changes in the atomic structure that led to different elements—a trait that escaped both models (Carazza and Robotti 2002, 305).

The experimental results on which Thomson's and then Rutherford's models were built had to be adequately connected with the results of experiments with the radiation spectra to produce a more adequate model. Neither Rutherford's model nor Thomson's was concerned with the results and intricacies of spectroscopy, nor could either model tally with the lower hypotheses generated from spectroscopic experiments, except in a rough qualitative sense. Rutherford and Thomson's experimental designs provided the key results which they then extended to their own models, but these two physicists probed a limited scope of the experimental domain, and their models could account only for particular aspects of the entire experimental context. In Bohr's hands, Rutherford's attempted master hypothesis that assimilated Thomson's concept of positive and negative charges distributed in the atom but incorporated the nuclear positive charge and orbiting electrons—thereby "transferring Thomson's techniques to the nuclear atom" (Helibron 2013, 26)—became a useful intermediary hypothesis, connecting a crucial domain of the experimental context with the master hypothesis Bohr devised.

In his 1925 paper that paved the way for his complementarity principle, Bohr basically summarizes all of this. He details (1925, 846) the experiments with spectral lines, in which *"the simple empirical regularities* [author's italics] among the spectral frequencies" (i.e., patterns of spectral lines) were drawn from experimental particulars. He then notes that Lenard tried to square these new lower hypotheses with the notion of electrons drawn from experiments with electric discharges of gases and X-rays. He offers a compressed characterization of the production of experimental, intermediate, and master hypotheses, as they are developed with respect to the regularities of the spectral lines:

This view [on the nonmechanical nature of the atom] is in general conformity with the spectroscopic evidence. An important feature of this evidence is the discovery of Rydberg, that in spite of the more complicated structure of the spectra of other elements compared to that of hydrogen, the same constant as that in the Balmer formula appears in the empirical formulae of the series spectra of all lines (Bohr 1925, 849).

This result is an autonomous lower-level experimental hypothesis drawn from experimental regularities. Now an intermediate hypothesis (i.e., of Conway and Ritz) is formulated, whereby "this discovery is simply explained by regarding the series spectra as evidence of the processes by which an electron is added to an atom, its binding becoming more firm step by step with the emission of radiation" (Bohr 1925, 849), instead of an atom as a whole producing spectral lines. Finally, the "step-like" intermediary hypothesis (i.e., the notion of "terms") is then assimilated in terms of orbits in Rutherford's master hypothesis of the atomic model, and then Bohr's.

*

Bohr was acutely aware of the importance and danger of taking hypotheses that were adequate in a certain domain of the existing experimental context too far or, alternatively, of failing to take them into account as intermediary hypotheses in light of their partial adequacy. This was the methodological rule he followed in inducing his model of the atom, and later on in formulating his complementarity principle. Experimentation is a demanding and arduous process that continues until the experimental results elicit a hypothesis that is acceptable across different experiments. Therefore, the process of experimentation ought to be as thorough as the resources and the experimenters' imagination allow. It should surround and examine the relevant "piece of nature" from all sides, as it were, and open it to the experimenter's senses through the most thorough manipulation possible.

Bohr's skill at putting together a more general hypothesis from lower ones in a clear and demonstrable way was his main tool in devising his breakthroughs. In that process, opting for particular metaphysical principles and concepts that the physicist deems fundamental for understanding the physical phenomena at stake and formulating the master hypothesis (the basic concepts thought essential for devising the "language" of the hypothesis) must be kept tightly in check by, and selected only with the guidance of, the lower and intermediary hypotheses. They must be bound by the demand for the comprehensibility of the master hypothesis and its ability to encompass the relevant experimental context. Bohr always prioritized the coordination and assimilation of partial accounts of phenomena devised by others from lower-level hypotheses into a general hypothesis over particular preferences for metaphysical soundness and mathematical beauty of the latter.[12] As Rosenfeld commented:

> He would dismiss the usual considerations of simplicity, elegance or even consistency with the remark that such qualities can only be properly judged after the event. . . . What he would rather insist upon, in discussing a proposed conjecture, was whether we could draw arguments in favour of it from the available evidence: thus logical analysis was not for him a mere verification of consistency (which he regarded as trivial), but a powerful constructive tool, orienting the groping mind in the right direction (Rosenfeld 1968, 117).

The demand for consistency was not something Bohr trivialized in order to push his (alleged) philosophical agenda, but something he felt had to be balanced with relevant empirical reasons. A careful bottom-up construction of hypotheses and adequate concepts to provide upward synthesis was his priority, and it was bound to be at odds with ready-made metaphysical preferences and common intuitions. Although the widely satisfying extent of the ontological coherence of the account is one thing which might eventually emerge from gradually connecting experimental particulars into an ever more general hypothesis, even a more moderate demand for overall consistency was secondary, as it had to be supported by the basic requirement of experimental comprehensiveness—that is, the careful inclusion of all relevant experimental particulars.

As a modeler, Bohr focused on the coherence of particular aspects of the model via intermediary hypotheses built on relevant lower-level hypotheses. At that stage of the development of the theory, this meant sacrificing the wholesale internal consistency of the master-hypothesis model to retain certain preferred simple general principles.[13] Bohr was aware what exactly he was sacrificing; for example, he labeled his synthesis of the classical planetary system with quantum conditions, to which we will turn shortly, as a synthesis with "horrid" assumptions (Darrigol 1992, 89). He also knew what was he gaining by such an approach: a thorough, precise, and usable agreement with the lower-level hypotheses. Thus, for instance, the model lacked a basic unity, another "major regulative ideal in physics," as Michel Bitbol (2017, 48) correctly states. Yet such a demanding ideal, along with others, was confined to a certain stream in the history of physics.

The price of all this was that of being accused of nonintuitiveness, or even obscurantism of resulting master hypotheses, by those physicists who based their view of quantum microphysical states and processes on a very particular metaphysical view as a necessary precondition for understanding any physical states, including quantum states. Henri Poincaré was an early critic of that sort (Heilbron 2013, 28). In 1911, he leveled this type of criticism at the physicists at the Solvay Council who were working to bridge quantum and classical physics piecemeal. He labeled them conceptually irresponsible as they, as he alleged, introduced contradictory principles that enabled inference of any desired result. Bohr thought of this remark as being mathematically driven, naive, and methodologically biased (ibid.). He voiced a similar criticism of the young physicists who were trying to establish the mathematical foundation of quantum electrodynamics a decade and a half later. To return to Poincaré's point, it was pretty clear for Bohr that experimenters did not opt for logical contradictions just like that, but dealt with experimentally based discrepancies (or paradoxes, as Heisenberg labeled them much later) to be synthesized; and this demanded trying out a variety of explorative provisional hypotheses that fell short of Poincaré's impatient imposition of a high-minded criterion.

As Helge Kragh correctly states, "From a methodological point of view, Bohr's theory was markedly eclectic, relying on a peculiar mixture of empirical evidence and theoretical reasoning" (Kragh 2013, 9). Physicists like Poincaré would certainly find this peculiar. The emergence of this methodological foundation was already apparent in Bohr's PhD dissertation (Bohr 1972–2008, vol. 1). And in his memorandum to Rutherford that preceded the 1913 papers that made him famous, Bohr called for the "exploration of the whole group of experimental results" which supported the Planck-Einstein hypothesis (Bohr 1972–2008, vol. 2, 136). This "peculiar" sort of exploration was, in fact, ongoing; "'quantum theory' in 1912 was a hodge-podge of postulates among which atom modellers could choose what best suited their needs" (Heilbron 2013, 29).

*

At a higher level of hypothesis construction, the requirement of a comprehensive grasp of the phenomenon of interest aims to prevent the susceptible intellect from being guided by conveniently chosen sensations and particulars. The experimenter is prone to such partial inductions if she confines herself to a preferred limited experimental context with a restricted grasp. Generally speaking, discrepancies across lower hypotheses will often present themselves to experimentalists early on, and will be reinforced in the process of induction of higher-level hypotheses the way they were reinforced in spectroscopy prior to Balmer's and Rydberg's work, or in the needle account of radiation.[14] When sorting out the experimental context, Bohr had to take into account this possibility of a reinforcing feedback loop between the way sensations are directed to choose experimental particulars and the language used to formulate lower hypotheses. The formulation of opposing hypotheses stemming from different experiments can invoke the selectivity of the experimenter's senses by emphasizing some aspects of observations (particulars) as mere appearances while singling out others as those that adequately capture the actual physical process. Often the number of experimental particulars across experiments is simply too large to lead unambiguously and effectively to adequate interpretation and understanding.

The question that the physicists in the role of mediators and moderators, a role in which Niels Bohr excelled, try to resolve is not whether our senses operate partially when co-opted by theoretical preferences—they are quite certain that they do—but in what precise manner they are partial within a given experimental context. This is why Bohr worked closely with the experimenters all along. A second stage of induction could commence only after a thorough understanding of the experiments was obtained. Bohr would never try to hastily "outline any finished picture, but [would] patiently go through all the phases of the development of a problem, starting from some apparent paradox and gradually leading to its elucidation" (Rozental 1968, 117). An exceedingly high level of vigilance is required when the first and the second stage of the inductive process come together, and hypotheses are either defined as intermediary or abandoned.

In 1913, obtaining a comprehensible grasp of the experimental context—that is, of all relevant experimental results—was a demanding task. The lower hypotheses based on various experiments could not be easily reconciled, and they elicited discrepant and seemingly contradictory views. John L. Heilbron and Thomas S. Kuhn (1969, 210) state that "a concern for *detailed* atom models was not widespread" at the time, although the participants of the Solvay meeting agreed that the way forward was analysis of the atomic structure and its inner workings (Aaserud and Heilbron 2013, 150; emphasis added). Modeling of atoms was widespread and was pursued by Thomson, Rutherford, Jeans, John W. Nicholson, and others; but the models were somewhat tenuous, aiming at general qualitative agreement with lower hypothesis. Sommerfeld's attempt to reconcile electromagnetism with Planck's theory of radiation never aimed at the full-blown model; rather, it was the result of the treatment of particular problems. Sommerfeld "derived results replacing theoretical analysis as he avoided any modelling of intra-atomic process" (Seth 2010, 150); or, to put it differently, he kept his theoretical ambition at the level of adequate intermediary hypotheses as most physicists, like those working on spectroscopy, did at the time. Bohr's model of the atom aimed at something much more daring: a precise quantitative agreement with the relevant sets of lower experimental hypotheses. He took full advantage of the capabilities of the inductive/modeling approach to physical phenomena.

Two supportive intermediary hypotheses induced from the lower hypotheses provided the initial foundation for Bohr's emerging model. The first one, based on Rutherford's experiments, was that there was a pointlike nucleus of the atom, with electrons orbiting in shells. This hypothesis agreed with the experimental results on ordinary radiation phenomena tied to electrons, as well as radioactivity. In fact, Rutherford's nuclear model could make sense of the distinction between radioactive phenomena due to the activity of the nucleus, and ordinary radiation for which electrons were responsible (Heilbron 2013, 30). Conversations with Hevesy on the structure of the atom and the weights of chemical elements were crucial for Bohr's understanding of this point; and, contrary to Rutherford's skepticism and his rather vague account of the inner atomic structure, Bohr saw the nucleus as the source of both alpha and beta particles (Aaserud and Heilbron, 2013, 162–63). The second intermediary hypothesis suggested that, though electrons are distributed in shells around the nucleus, they must move within *a stable atomic structure*. The ramifications and extent of this stability were induced from spectroscopy—that is, from the discrete spectral lines produced by gases.

The limitation of these two hypotheses taken together was that an adequate model of the atom had to overcome the discrepancy (paradox) whereby there cannot be a *mechanical* stability of the atom. In trying to calculate the values of electron orbits, Bohr discovered that perturbed vibrations of electrons at the equilibrium orbit are not stable mechanically, and that a particle passing by the atom will result in its demise.[15] He immediately informed Rutherford of this discovery, understanding its immense implications and, eventually, "in his characteristic way of reasoning, made a positive feature [of it] and coupled [it] to the further demerit that the laws of mechanics do not determine the size of the rings of planetary atoms" (Aaserud and Heilbron 2013, 164).

Bohr succinctly summarized these hypotheses and laid out the challenge his model faced. "It is evident that systems like the nuclear atom"—the first intermediary hypothesis—"if based upon the usual mechanical and electrodynamical conceptions"—the limitation due to the second intermediary hypothesis—"would not even possess sufficient stability to give a spectrum consisting of sharp lines" (Bohr 1922a, 21). Thus, for Bohr, renouncing mechanical principles was a collateral loss accompanying his attempt to avoid a paradoxical picture of the atom based on mechanical presuppositions—paradoxical in light of the experimental results (sharp spectral lines implying a stable inner atomic state)—and it became part of the induction of a hypothesis that encompassed the main elements induced from the experiments.[16] Bohr was ready to pay the price of abandoning the intuitive and traditional appeal of mechanical principles for the experimental comprehensiveness and plausibility of the model in light of experimental hypotheses.

Rutherford's nuclear model was an attempt at a master hypothesis, but it faced numerous paradoxes in light of the existing experimental context. As noted previously, the atom could not be stable, as mechanical forces could not keep an electron rotating continuously around the orbit. In addition, the size of the orbit, the exact radius of the atom, could not be determined on the basis of the time of vibrations. The value of vibrations did not distinguish between various radii, and thus could not provide an argument for stability. Moreover, Maxwell's classical theory of radiation allowed all possible values of electron orbits, thus creating the possibility of the continuous dissipation of the energy of the rotating electron, the continuous increase of the frequency of the electron's revolution, and the continuous decrease in the length of its orbit.[17] Yet, contrary to these classical predictions, the above-mentioned experiments with spectra demonstrated that atoms are fairly stable systems of a finite size, and that they radiate light in discrete packages. Bohr was seeking a unifying mechanism behind this.

Bohr addressed deficiencies of Rutherford's model iconoclastically, by abandoning the classical mechanical route for handling discrepancy. This seemed to him the only viable option for reconciling discrepancies in the overall experimental context. The model could fit the actual experimental context, and could avoid paradoxes only if it was "a hypothesis for which there will be no attempt at a mechanical foundation" (Bohr 1922a, 136). For Bohr, everything pointed in that direction, and it seemed "to be nothing else than what was to be expected as it seems rigorously proved that the mechanics cannot explain the facts in problems dealing with single atoms" (ibid.).

This in fact confirmed the adequacy of the possibility that Bohr had contemplated when he was writing his PhD dissertation and thinking about various experimental results bearing on the nature of the atom. An insight he explicated in his PhD dissertation unequivocally affirms the priority of the experimental context in the elicitation of hypotheses over even the most cherished ontological preferences. He stated, "The assumption [of mechanical forces] is not a priori self-evident, for one must assume [i.e., by recognizing the relevant experiments, especially the difficulties in accounting classically for the magnetic properties of metals] that there are forces in nature of a kind completely different from the usual mechanical sort" (Bohr 1972–2008, vol. 1, 175). Bohr already suspected that a completely new set of postulates might have to be constructed to encompass all other relevant experimental results. Yet, despite this strong formulation, Bohr was aware that a much more detailed study was needed to determine whether this was the case, and to develop a preliminary postulate. This was still the conjectural insight of a young physicist on the current state of the field, a far cry from a comprehensive master hypothesis that could overcome "deficient concepts" by building new ones via multiple, seemingly discrepant lower hypotheses.

Bohr made the first step on this road in his dissertation, demonstrating, in the spirit of Planck, that rigidly following the principle of the equipartition of energy (equal partition of energy across oscillating particles that radiate), which Planck had abandoned in his quantum version of the radiation law, cannot account for the effect of external magnetic fields on bound or free electrons in metals. This result was also pointed out at the Solvay Council in 1911, and it was perhaps the main achievement of Bohr's dissertation. In fact, this part of his work makes it quite apparent, perhaps more than anything else, that Bohr had no preconceived model in mind (Heilbron and Kuhn 1969, 213–18) but worked his way to the model from the bottom up. The model was built very gradually, as was the choice of various elements, including relevant experiments and experimental laws (the lower hypotheses). Only after Bohr updated his knowledge in Thomson's and Rutherford's labs among a wider community of experimentalists could he work out the details of a nonclassical model by inducing it from the overall experimental context.

Heilbron and Kuhn state that the reason guiding Bohr to develop his model lay "not in the general conviction of the need for quantum theory which Bohr drew from his theses research, but rather in certain *specific* problems with which he busied himself until almost the end of his year in England" (1969, 212). This is generally correct, but it unjustifiably and perhaps unintentionally downplays the confidence Bohr had in the lower-level experimental hypotheses. It is true that his general impression that a new theory was needed was not a prior independent motif antecedent to his research; it gradually sprang from his consideration of a number of such hypotheses and the discrepancies they generated.

In fact, even at the Solvay Council, a few most prominent physicists working on the quantum account of radiation started subsuming the old physics that could not be reconciled with new experimental results under the label "classical physics" (Heilbron 2013, 27). The list of such relevant experimental results was long (Aaserud and Heilbron 2013, 144), and it sharpened the distinction between quantum radiation theory and old mechanics already established by the failure of the equipartition principle. Einstein called explicitly for a replacement of "classical mechanics" (ibid., 146). Nor was the idea itself new to physicists; the so-called energeticists, a group of scientists including Mach, had been arguing for the abandonment of a mechanical account of the physical world, albeit for different reasons (Seth 2010, 143–44).

In contrast, a group of physicists from Cambridge, including Thomson, Lord Rayleigh, and Jeans, made a last desperate attempt to escape the implications of Planck's quantum and save the equipartition approach they deemed physically inescapable. They introduced the notion of delayed equipartition that could take millennia to actually occur (Heilbron 2013, 27). This was a typical instance of a mathematically driven approach in the physics of that era, a search for an elegant mathematical tool or technique that could serve as a silver-bullet solution to the burning dilemmas. But this sort of approach bore fruit only in the later development of quantum theory toward quantum mechanics, with Heisenberg, Dirac, and others. For the time being, it was defeated by the steady march of partial and mediating hypotheses that grew directly out of lower experimental hypotheses. Poincaré explicated this fact at the Solvay Council meeting, as a criticism of the above-mentioned suggestion of the Cambridge school, Jeans in particular. He stated that theories "must establish a connection among the various experimental facts and, above all, support prediction" Langevin and de Broglie 1912, 77) instead of postulating arbitrary constant parameters for observed phenomena. Yet, anticipating an argument against quantum jumps in Bohr's model that Erwin Schrödinger would make a decade later, Poincaré also expressed skepticism about the discontinuities that seemed inevitable if Planck's approach to radiation and understanding of the atom were followed (Heilbron 2013, 28). The oscillating electron within the atom could not change its state continuously if Planck's law were followed. This meant that differential equations could not be applied in accounting for it. This was precisely the kind of mathematical formalism that Schrödinger deemed central and would use in the intuitive understanding of microphysical states a decade and a half later.

During his stay at Thomson's laboratory at Cambridge, Bohr met and conversed with S. B. McLaren (Rozental 1968, 42), a lecturer at Birmingham who published a paper (McLaren 1913) arguing that any attempts to account for radiation phenomena within the classical framework would fail. The key to the success of such attempts was the adequate introduction of Planck's constant, the argument went. Bohr visited Birmingham to exchange ideas on this matter with McLaren, as physicists who shared this particular view were rare, bar a few physicists at the Solvay Council. Bohr also must have been updated on the latest developments in the quantum theory of radiation initiated by Planck, as these were discussed at the Solvay Council in Brussels. Rutherford, in whose laboratory Bohr worked at the time, was selected as a prominent participant in the council session, and other physicists from the laboratory attended.[18] Thus, Bohr took Planck's treatment of radiation as a third and major supportive intermediary hypothesis that connected the lower hypotheses on radiation synthesized by Planck's law of radiation with the main features of the overall model. Planck himself tried applying quantization to energy, as well as to the action of the atom (Seth 2010, 156–62).

Planck's hypothesis suggested a way to reconcile the first two intermediary hypotheses: the nuclear hypothesis along with the electron orbits, and the hypothesis about stable atomic structure. Accordingly, Bohr introduced his *quantum postulates*, defining the main features of the model that resolved the discrepancy between the first two intermediary hypotheses generated by the classical mechanical approach. Heilbron succinctly summarizes the way Bohr made use of Planck's hypothesis as a third intermediary one:

> Bohr looked to the quantum hodge-podge for a suitable rule. He found it in a strained analogy to Planck's restriction on the oscillators at the heart of his radiation theory: in their "permanent" or ground state, achieved after they have radiated away all the energy that nature allows them to dispose of, every electron bound in a nuclear atom, regardless of the radius of its ring, has a kinetic energy T equal to K times its orbital frequency (2013, 30).[19]

It was in fact a long grueling process of aligning details of Planck's law, its latest installment—that is, Planck's "second theory"—with various aspects of the model as characterized by the first two intermediary hypotheses

(Heilbron 2013; Kragh 2012). Prior to that, Bohr had used Planck's laws in his PhD dissertation to offer an account of magnetism more adequate than the one proposed by Hendrik Lorentz, which worked only for ensembles of electrons, not for single ones. This pushed Bohr's research in the direction of the structure of the atom.[20]

Planck's "second theory" postulated the discontinuity of absorption and the classical nature of emission in Planck's resonators. In Planck's words, he "located the discontinuity at the place where it can do the least harm, at the excitation of the oscillator" (Planck in Kuhn 1987, 236).[21] A central idea Bohr gradually arrived at, however, was "to equate the spectral frequency n with the average orbital frequencies involved when the bare hydrogen nucleus captures an electron into the n-th 'stationary state'" (Heilbron 2013, 44). This quantization of the relation between an electron's kinetic energy and its orbital frequency provided *certain* stable orbits; a constant defining such orbits (eventually identified with Planck's constant) was equal to the ratio of the former and the latter. These rather ad hoc attempts to relate orbital frequencies and kinetic energy were tried in other domains such as magnetism or photo-effect (Heilbron and Kuhn 1969, 244). But Bohr's model provided a stunningly precise agreement with the lower level hypotheses.

Thus, first, electrons rotate only along such "privileged" orbits. The overall work of the electron must be an integer multiple of h: $J = nh$ (quantum number $n = 1, 2, 3 \ldots$). Second, radiation is not released if an electron moves around an allowed orbit; it occupies a stationary state, and the atom is stable. Electrons can spontaneously move from a higher to a lower orbit; the difference in the energy between the two is released as radiation of the frequency proportional to that difference, or they can move to the higher orbit if the atom absorbs energy in the collision. The frequency of the radiated or absorbed energy is determined by the difference in quantized energy states of respective orbits: $v = (E1 - E2)/h$. This frequency determines both the mechanical energy of the rotating electron and the optical frequency of radiation.[22]

Planck's account had assumed a harmonic oscillator that vibrated at the same frequency in all energy states. Bohr's modified version got rid of this assumption, effectively removing the direct link between vibration frequencies and energies and opening the door to experimental results, most notably those in spectroscopy (Heilbron 2013, 46). If the radiation of plausible frequencies had not matched experimental findings, Bohr's model would remain one in a series of similar valiant attempts to develop an adequate master hypothesis, like those of Nicholson or Jeans.[23] And as such, it would have been only nominally more ambitious than Sommerfeld's attempted synthesis of electrodynamics based on experiments with polarization on the one hand, and Planck's law on the other. Sommerfeld actually pursued a project very similar to Bohr's, yet without the matching ambition and vision of a master hypothesis, an all-encompassing model of atomic internal structure and atomic interactions (Seth 2010).

The successful reconciliation of the stability of the atom and the electrons orbiting around its nucleus required the refinement of those intermediary hypotheses that connected spectral experimental laws, built on the lower hypotheses of Balmer and Rydberg, and the quantum features of the model. The fact that the quantum hypothesis was not a topic of interest beyond the circle of physicists concerned with radiation alone makes this leap even more impressive. Bohr's continuous wide "scanning" of the experimental work via his network of colleagues was unusual, but crucial. Hans M. Hansen, one of the experimenters working with spectral analysis, told Bohr of the existence of the Balmer lines, the distribution of which satisfied the numerical Rydberg rule (Bohr 1962; Pais 1991, 144) and turned out to correspond to the distribution of quantized orbital energy stages in Bohr's model. This enabled Bohr's assimilation of spectroscopic experimental results in his atomic model. Outside spectroscopy, both Rydberg's formula and Balmer's lines were seen as merely curious patterns with no deeper physical significance, especially not for building full-blown models of the atom. As Bohr put it, "They were looked upon in the same way as the lovely patterns on the wings of butterflies; their beauty can be admired, but they are not supposed to reveal any fundamental biological laws" (Bohr in Rozental 1968, 51). In fact, Bohr at first explicitly told Rutherford that he did not plan to deal with spectral lines of intensity (ibid., 52). Yet Hansen's remarks immediately struck Bohr as a crucial lower hypothesis. Hansen pointed out to Bohr the significance of the hypotheses to the field of spectroscopy, crucially including the notion of "terms," introduced by Conway and Ritz (Heilbron 2013, 43).[24] Without this, Bohr's treatment would not have exceeded the quantitative and qualitative adequacy of Nicholson's model which, interestingly enough, could have been revised in the same direction. Bohr's initial model, prior to its assimilation of spectroscopy, was actually a simpler model of that sort. This crucial step of relating it with spectral analysis was the basis of the model's most useful implications. It demonstrates that the success and focus of Bohr's model relied on its agreement with the lower-level hypotheses, and this far outweighed its failure to provide an ideally consistent "intuitive" picture.

Bohr had gradually become an unusually ambitious modeler, certainly more ambitious than Thomson or even Rutherford. When he started working on spectroscopic results, "he expected to achieve a coherent new position which, expressed in consistent models, would yield results in exact quantitative agreement with experiment" (Heilbron and Kuhn 1969, 226). Bohr expected that a model of the atom could be elaborate enough to constitute an adequate master hypothesis, comprehensive enough to provide a quantitatively precise agreement with the lower-level hypotheses. This was in stark contrast to Thomson, who did not seek quantitative agreement between the data and his models; to Poincaré, who aimed at models as rounded mathematical devices; and to Sommerfeld in his phenomenological phase, when a master hypothesis was supposed to be a direct inference of theoretical regularities from experimental phenomena and laws. To succeed, this extraordinarily ambitious approach required an extraordinary agreement with the experimental results.

Thus, on the the basis of his model's quantum feature, Bohr famously derived the Balmer formula for the spectral series of hydrogen, given the experimental values of an electron's electric charge, its mass, and Planck's constant. This was the decisive step in constructing a comprehensive model in close agreement with the experimental context. In fact, "up to that time no one had ever produced anything like it in the realm of spectroscopy, agreement between theory and experiment to five significant figures" (Pais 1991, 149).

Yet the step was not simply the application of a ready-made model. Rather, as Heilbron states, and as Heilbron and Kuhn (1969) have demonstrated in detail discussing a three-part paper Bohr wrote, "the order of publication of the parts thus hides the order of their conception: the applications to the 'constitution of atoms and molecules,' to use the title carried by the entire sequence, did not extend the principles apparently established by agreement between theory and measurement of hydrogen's Balmer lines, but antedated the systematic consideration of spectra" (Heilbron 2013, 39). In this way, inductive steps proceeded back and forth from suitable applications to the refinements of the model: from lower to intermediary to master hypotheses, and then back by using the satisfying results achieved in the process.

Nicholson, in fact, constructed his own model before Bohr constructed his. Like Bohr, he (Nicholson 1911) tried to amalgamate the nuclear hypothesis with quantized values of radiation. He postulated *uratoms*, whereby energy states of electrons differed discontinuously and could radiate sets of spectral lines—clearly a legacy of Ritz and Conway as well. The models of Nicholson and Bohr represented two very similar attempts at a master hypothesis, but Nicholson's treatment of spectral laws was not as subtle and as successful as Bohr's. It could not account for the actual spectral lines, especially not with the precision of Bohr's model, nor was it intended to do so (Heilbron 2013, 42–43). Bohr understood that, based on Rutherford's experiments with the alpha particle absorption, and contrary to Nicholson's interpretation, the behavior of alpha particles indicated that an atom could exist with a single electron. This understanding proved to be far-reaching, as alpha particles were still poorly understood. Darwin's (1912) calculations of the loss of velocity of alpha particles, built on the results of experiments agreeing with the values predicted by Rutherford's model, were instrumental for Bohr's model of the structure of the atom. Bohr questioned Darwin's results, seeking a more precise agreement with the experiments while rejecting the classical assumptions on the nature of the interactions between the atom and alpha particles (Heilbron and Kuhn 1969, 239–40).

In his groundbreaking three-part paper "On the Constitution of Atoms and Molecules," Bohr (1913a, 1913c, 1913d) successfully applied Planck's quantum of action to Rutherford's classical model consisting of a nucleus at the center of elliptically orbiting electrons. Accommodating the experimental context, he postulated the existence of stationary orbital states: electrons may occupy only certain, viz., stationary orbital states, thus providing for the stability of the system. Bohr described his model "as the only one which seems to offer a possibility of an explanation of the whole group of experimental results, which gather about, and seems to confirm perceptions of the mechanisms of the radiation as the ones proposed by Planck and Einstein, —that is, via quanta" (Bohr 1972–2008, vol. 2, 136; figure 5).

Exactly how large the entire group of experiments was that the model, including its later instances, successfully accounted for is an open question. The most immediate domain of the first version of the model as the master hypothesis is clear, however, as that domain played a major role in establishing the hypothesis. Bohr's pronouncement was certainly true regarding the existing spectral lower hypotheses that his model cemented as experimental laws to be taken into account in any future development of models of the atom. But it is not clear how far the truth of Bohr's statement could be extended into the molecular domain—especially because over the years, and until the emergence of quantum mechanics, the model would

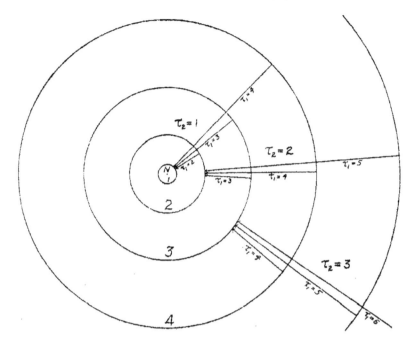

Figure 5. Probably the first published depiction of Bohr's model of the atom. The nucleus (N) of a hydrogen atom is positioned at the center, and the orbits of electrons are represented as rings designated by integers. Various "drops" of electrons from higher to lower orbits, designated as τ, correspond to the occurrences of Balmer, Ritz, and Paschen spectral lines.

From W. D. Harkins and E. D. Wilson, "Recent Work on the Structure of the Atom," *Journal of the American Chemical Society* 37, no. 6 (1915): 1396–1421. Copyright 1915 American Chemical Society. Reprinted with permission.

be reformulated by Bohr and other physicists, especially Sommerfeld.[25] Precisely because of the inductive nature of the process that produced the model, the domain cannot be fixed in any precise way; only various areas of experimental work can be graded in terms of the precision of their agreement with the model via intermediary hypotheses.

5: THE CORRESPONDENCE PRINCIPLE AS AN INTERMEDIARY HYPOTHESIS

Some intermediary hypotheses were the result of failed attempts at producing a master hypothesis that turned out to be adequate for only a certain domain of the experimental context. Like Thomson's and Rutherford's models of the atom, they reached too far, but certain aspects remained adequate. They were not completely abandoned in further pursuit for a viable master hypothesis, for example, the way Lenard's model or Jean's theory of equipartition were almost immediately abandoned by the community. Others were turned into useful intermediary hypotheses, each serving as a partial support of an emerging master hypothesis after being transferred successfully to a new domain, as in, for example, the way Bohr used Planck's law of radiation in his model.[1]

Some intermediary hypotheses were conceived as intermediary from the very beginning. They varied in scope, depending on the portion of the experimental context they aimed to grasp. Conway and Ritz formulated their hypotheses as explicit attempts to challenge the existing assumption about the atom as an oscillating whole. They connected the lower hypothesis in spectroscopy with a general feature of the atom's structure—a particular assumption upon which a whole set of models can rest—rather than with a full-scale model of the atom. Similarly, in the 1920s Samuel Goudsmit and George Uhlenbeck formulated a classification of spectral selection rules and energy levels without referring to or making use of any particular model of the atom (Kragh 1985, 114).

The history of spectroscopy also has a messy back-and-forth relationship between intermediary and lower-level hypotheses. The controversy over the fine structure of spectral lines mentioned earlier was eventually resolved with the gradual advent of the higher-level hypotheses of Ritz and Conway that assimilated the "works" by introducing the atomic "terms" as source of oscillations. But this was a bumpy road. Closely guided by Sommerfeld's theory, for instance, Paschen accounted for the details of the fine structure of the helium molecule (Paschen and Back 1912),[2] recognizing a separation of a doublet (he labeled this the IIIa line). Yet Paschen's classification was questioned, as some experimentalists doubted that this particular line was part of the ionized helium spectrum, as Paschen claimed; they thought it likely to be a result of excited He_2 molecules. In fact, "the comparison between theory and experiment"—or, rather, between intermediary hypotheses and lower-level hypotheses on the distribution of lines—"depended on the choice of intensity rules" (Kragh 1985, 93). Another line was predicted by a theory of Sommerfeld and Kramers but not sought for in fine-structure spectroscopy for a long time; its absence finally was an argument to abandon the theory. Only the gradual advent of quantum mechanics in the 1920s adequately and fully resolved these discrepancies and offered a unifying view of lower-level hypotheses.[3]

The correspondence principle, however, was an ambitious intermediary hypothesis, essential in developing and refining Bohr's full-scale model of the atom, and connecting it with the lower-level hypotheses. Kragh points out that the correspondence principle (CP) permeates Bohr's entire theory of the atom in an opaque way (2013, 9). But understanding that it was a versatile and constructive intermediary hypothesis that did not uniformly support the model as the other three intermediary hypotheses did should make it substantially less opaque.

The seeds of the principle were present in an idea of 1913, and perhaps even earlier (Heilbron 2013, 45). It was, in any case, a principle developed gradually by trial and error (ibid.). Bohr suggested treating frequencies of transitions between orbits of high n in the low frequency limit as analogous to the classical mechanical frequencies of an electron (i.e., to the states in the Fourier series analysis of motion) and thus to the frequencies of radiation calculated with such classical theory. Until 1918 his concern was the analogy with frequencies, not the more formal correspondence he gradually developed (Kragh 2012, 198). Initially, he used the label "analogy principle," and only introduced the notion of the "correspondence principle" in 1920 at a lecture in Berlin (Bohr 1972–2008, vol. 3, 241), although it would be hard to tell when exactly the analogy became a sufficiently versatile constructive hypothesis.

The use of analogies as powerful constructive tools was not typical only of Bohr's work. His approach was not "nonconventional" in that respect, as Bitbol (2017, 49) claims it was; nor do we need to probe the depths of Kantian philosophy to understand how "to bring seemingly heterogeneous theoretical structures into a unique system" the way Bohr supposedly uniquely did. In fact, Heisenberg's uncertainty principle was a hypothesis of that sort, albeit with a somewhat different goal: defining the exact limits of the supporting hypotheses. And Schrödinger developed his wave account by devising and examining Hamilton's optical-mechanical analogy (Joas and Lehner 2009), effectively turning it into a constructive intermediary hypothesis.

The CP was, in fact, a result of gradual bottom-up hypothesis-building from the experimental context within the confines of the model. A number of experimental results on the so-called frequency conditions that established the relationship between frequency and emitted light were essentially classical. Bohr understood that this insight had to be absorbed into an account of microphysical states predicated on quantum conditions, and he set out to develop an intermediary hypothesis starting with a substantial analogy. He noted the existence of a "far-reaching correspondence between the various types of possible transitions between the stationary states on the one hand and the various harmonic components of the motion on the other hand" (Bohr 1922a, 24). He stated its ultimate goal: "We can see it is possible to develop a formal theory of radiation, in which the spectrum of hydrogen and the simple spectrum of a Planck oscillator appear completely analogous" (ibid., 29). Reflecting on the role of the principle, he said we needed experimental facts "to develop our deficient theoretical concepts" (Bohr 1972–2008, vol. 3, 397). Although these classical theoretical concepts are deficient in that they cannot be used to formulate the master hypothesis, they can be usefully employed in the overall account of microphysical states via an intermediary hypothesis: "Considering transitions corresponding to large values of n' and n'' we may therefore hope to establish a certain connection with the ordinary theory" (Bohr 1922a, 26).

This idea of borrowing concepts of "old mechanics" to capture certain phenomena and combine them with the intermediary analogies with Planck's laws was already part of the initial stage of working out the model of the atom (Heilbron 2013, 46). Bohr also encountered the idea of fruitfully using the concepts from mechanics as analogies rather than as complete accounts in Joseph Larmor's work at Cambridge (Aaserud and Heilbron 2013). It turned out that it was possible to satisfyingly assimilate novel experimental particulars on spectral phenomena via such an intermediary hypothesis. The CP, still close to the experimental context, used known classical concepts to usefully bridge the experimental particulars (the lower hypotheses of spectroscopy) and the master hypothesis (the atomic model involving the quantized state transitions). The principle was gradually developed all the way to the final refining of the model to accommodate the spectra—that is, to determine how exactly this could be done in terms of the intensities and polarization of the spectral lines— functioning as one of the adequate "selection principles" (Kragh 2012, 199) and as formalization to match the lower laws quantitatively. The induction of the intermediary hypothesis (via "rational generalization," in Bohr's parlance) provided a useful tool for connecting an aspiring master hypothesis (the model of the atom) with a particular segment of the experimental context: "This correspondence is of such nature, that the present theory of spectra [based on his model] is *in a certain sense* to be regarded as a rational generalization of the ordinary theory of radiation" (Bohr 1922a, 24; emphasis added).

This goes to the heart of Bohr's method: it is a careful and gradual crafting of a central heuristic hypothesis, not a metaphysically or otherwise driven pursuit of models. In a memorandum to Rutherford, Bohr explicated his understanding that quantizing is a heuristic tool for a certain domain:

> It seems rigorously proved that mechanics is not able to explain experimental facts in problems dealing with single atoms. In analogy to what is known for other problems it seems however to be legitimate to use the mechanics in the investigation of the behavior of the system if we only look apart from the questions of stability (or of final statistical equilibria) (Bohr 1972–2008, vol. 2, xxiii).

As Bohr's key intermediary hypothesis, the CP was thus a bridge between the part of the experimental context that was inevitably expressed classically, and the quantum model of the atom: "This agreement clearly gives us a *connection between the spectrum and the atomic model of hydrogen*, which is as close as could reasonably be expected considering the fundamental difference between the ideas of the quantum theory and of the ordinary theory of radiation" (Bohr 1922a, 27; emphasis added). The ordinary theory of radiation accounts for the experimental particulars, while the quantum theory is an induced general (master) hypothesis, the concepts of the latter being detached from the terms that adequately describe the experiments. Thus, Bohr commented, "When the quantum numbers are large, the relative probability of a particular transition"—a key quantum feature of the quantum atomic model—"is connected in a simple manner with the amplitude of the corresponding harmonic component in the motion," as recorded in the experiments (ibid., 26). This statement was a direct result of the experiment with the Stark effect, thus bridging the experimental result with the model of the atom (ibid., 39).

In more concrete terms, what the CP grasped was that the results of transitions in the hydrogen atom model were due to the so-called Stark ef-

fect, a particular splitting of spectral lines due to the electric field. Rutherford pointed out to Bohr the potential agreement with his model (Bohr
1972–2008, vol. 2, 589). Bohr's formulation of this achievement was precise: "It is possible however to obtain a quantitative estimate of the relative
intensity of the various components of the Stark effect of hydrogen, by cor-
relating the numerical values of the coefficients . . . with the probability of
the corresponding transitions between the stationary states" (Bohr 1922a,
42). Bohr clarified this view:

> The correspondence principle suggests at once that these facts [transi-
> tions] are connected with the characteristic polarization observed in the
> Stark effect. . . . We would anticipate that a transition for which $(n' - n'')$
> $+ (k' - k'')$ is even would give rise to a component with an electric vector
> parallel to the field, while a transition for which $(n' - n'') + (k' - k'')$ is odd
> would correspond to a component with an electric vector perpendicular
> to the field. These results have been fully confirmed by experiment and
> correspond to the empirical rule of polarization, which Epstein proposed
> in his first paper on the Stark effect (ibid., 41–42).

This insight emerged gradually, however. The experiments with ionized
helium were an early indication of the analogous correspondence, as were
Henry Moseley's (1914) experiments with spectral lines. And in the case
of the conservation of the angular momentum, "we are led to compare
the radiation emitted during the transition between two stationary states
with the radiation which would be emitted by a harmonically oscillating
electron on the basis of the classical electrodynamics" (Bohr 1922a, 51).[4]

Moreover, the CP was indispensable for using the results to predict
relevant phenomena based on the atomic model: "It also enables us to
draw conclusions about the relative probabilities of the various possible
types of transitions from the values of the amplitudes of the harmonic
components" (Bohr 1922a, 52). In fact, in his 1922 piece on spectra and
atomic constitution, Bohr attempted to generalize the CP by assimilating
other various experimental results. He drew an explicit and sharp distinction between his discussion of his model of the structure of atoms and
molecules (the master hypothesis), on the one hand, and discussions of
spectral theory and the intermediary hypothesis that relates the model
to the experimental particulars, on the other (Bohr 1922, 59). He stated,
"With the aid of this general correspondence I shall try in the remainder
of this lecture to show how it is possible to present the theory of series
spectra and the effects produced by external fields of force upon these

spectra in a form which may be considered as the natural generalization of the foregoing considerations" (ibid., 37).

To sum up, first, the classical component of the principle is essentially nothing more than an analogy with the actual experimental particulars. Aaserud and Heilbron (2013, 177) acknowledge Bohr's use of classical mechanical formalism to formulate his result, while denying that classical mechanical laws fully apply: emphasizing the lack of full applicability "hedges" against the fact that Bohr's constant relating of orbital frequencies and frequencies of spectral lines was not yet directly linked to Planck's constant. Yet, rather than hedging, Bohr, in the spirit of what became his correspondence principle, may be simply pointing out the classical aspect of the lower-level hypothesis (Balmer and Rydberg's rules concerning spectral lines) in connection with clearly nonclassical atomic structure that produced the observed spectral lines. Second, it quite directly ties the experimental results—with their classical description—to the inevitably nonclassical properties of the model. Thus, the classical mechanical frequencies provide an adequate grasp of the experimental particulars, while the optical frequencies occur because of the orbital transitions at high n. In Bohr's words: "If we consider the harmonic component of the motion we obtain a simple explanation both of the non-occurrence of certain transitions and of the observed polarization" (Bohr 1922a, 43).

The CP is often understood as the mark of a sharp divide between the micro and macro worlds, treated as quantum and classical respectively, as if "the existence of the quantum h makes the physics of the microworld conceptually distinct, indeed unreachable, from that of the macroworld even where, as at the correspondence limit, the two systems give the same numerical result" (Aaserud and Heilbron 2013, 192). "Unreachable" may be too strong a term to account for the relationship between quantum theory and classical mechanics at that point in the development of quantum theory. It was certainly not analogous to the relationship between Newtonian mechanics and the general theory of relativity, where the former can potentially be embedded in the latter one way or another, the former being an approximation of the latter. But a sharp distinction between the two certainly was not a primary conclusion on the discontinuity of the macro- and microphysical world (as if microphysical and quantum are where h is) and definitely not a driving force of the CP.

First, in the CP, classical formulations are used as convenient means of generalizing lower-level hypotheses (defined by everyday language notions, and by classical-mechanical notions as their extensions) to build a general model, making these hypotheses "reachable" at the higher theoretical level of hypotheses. Second, Bohr was fully aware and even stated explicitly that discrete values of frequencies, monochromatic ones, do not really exist (Kragh 2012, 202). The classical values figuring in relevant lower-level laws are the result of experimental particulars; the theoretical model offers an adequate insight, but the experimental results confirming it are inevitably classical in nature because of the way our senses work: "In the limiting region of large quantum numbers there is in no wise a question of a gradual diminution of the difference between the description by the quantum theory of the phenomena of radiation and the ideas of classical eletrodynamics, but only of an asymptotic agreement of statistical results" (Bohr 1972–2008, vol. 3, 480). To think that Bohr's CP pronounced dichotomous kingdoms of classical and quantum is misleading. There was no space for a strong or a metaphysical dichotomy here. The classical and quantum domains are not on par with respect to inherent physical properties. Bohr stated forcefully that "in fact . . . the Correspondence Principle must be regarded purely as a law of the quantum theory" (ibid.). He reiterated this in his 1928 summary, stating that "this connexion cannot be regarded as a gradual transition towards classical theory" (Bohr 1928, 589). Yet it is an operational intermediary law or hypothesis. The CP is a law of quantum, not classical, theory, but this "can in no way diminish the contrast" or its heuristic value between the postulates"—that is, the quantum postulates of the atomic model "and the electromagnetic theory," a classical theory used to conveniently summarize and express experimental results.

Thus, attempts to seek the origin of the CP in a particular philosophical doctrine like that of H. Høffding's account of discontinuity (Seth 2010, 197) are unconvincing. The trajectory of its development and different phases of this development clearly show that we can hardly reduce its aim to mending classical continuity and quantum discontinuity (ibid., 198). In fact, such an aim was neither explicated by Bohr nor apparent in the development of the principle. It worked as a bridge between classical experimental reports and a quantized model of the atom. Otherwise it would have lacked the heuristic property needed to achieve what it had achieved, namely a lack of rigidity. As Heisenberg stated in an interview to Thomas Kuhn, "I always liked Bohr's Correspondence Principle just because it gave that kind of lack of rigidity, that flexibility in the picture, which could lead to real mathematical schemes" (Heisenberg in Kuhn 1963, 13).

Kramers suggested that the CP was not defined in a precise manner as a theoretical postulate in an easily applicable quantitative form (Holst et al.

1923, 139). And it was condescendingly labeled a magic wand by Sommer-feld (1921, 400), as well as a "bright spot" in "this night of difficulty and uncertainty" (Kramers 1923, 550).

The CP was certainly not a fragment of an overall account. As Kragh states, "The elementary recognition that the results of classical and quantum physics converge in the limit $h \to 0$ (or $h\nu \to 0$), sometimes known as 'Planck's correspondence principle,' does not capture what the correspondence principle, properly understood is about" (2012, 197). But it is also far more than a calculating device; Aaserud and Heilbron seem to suggest it was just that (2013, 192). It is an ever-expanding and ever-refining intermediary hypothesis aimed at experimentally substantiating the master hypothesis. And, as such, it is quite expectedly somewhat mystifying if we are looking for a definitive and precise formulation; as in various instances of Bohr's model of the atom, different versions of the principle, "logic and clarity apart, . . . were empirically fruitful" (Kragh 2012, 202). Nothing other than particular conceptual or metaphysical commitments could have prevented a careful bridging of quantum and classical domains this way, as long as it was rooted in the experiments.

Yet some physicists, notably Sommerfeld and Pauli, argued that this was an indication of a major deficiency of the principle. Sommerfeld stated that "without clearing up the conceptual difficulties it allows [Bohr] to make the results of the classical wave theory directly useful for the quantum theory" (1919, in Kragh 2012, 210). He argued that his collaborative account with Wojciech S. P. Rubinowicz was better on this point, though it turned out to be not as fruitful as the CP. He admitted defeat and acknowledged that the principle was heuristically irreplaceable (Sommerfeld 1921, 400), but he also sharpened his criticism, writing to Bohr: "I must confess that your principle, the origin of which is foreign to the quantum theory, is still a source of distress to me, however much I recognize that through it a most important connection between quantum theory and classical electrodynamics is revealed" (Sommerfeld in Kragh 2012, 210). This was in stark contrast to Bohr's view that the principle was an integral part of quantum theory. In a meeting of the Society of German Scientists and Physicians, Sommerfeld complained about the CP "mixing" quantum and classical viewpoints (Kragh 2012, 211), arguing that a satisfying theory should constitute a proper axiomatic-deductive system, and thus revealing a fundamental methodological difference driving his criticism. This view of the CP as "foreign to the quantum theory" and conceptually deficient is strongly prescient of the charges that Schrödinger laid against Bohr's complementarity account several years later. Sommerfeld also complained about murky conceptual grounds, emphasizing the intuitiveness of his own interpretation and claiming that it was based on solid conceptual grounds. The motivation for the criticism came from the physicists who valued conceptual coherence and the principled basis of a theory very highly (at least in that particular phase of Sommerfeld's work), and regarded them as almost inevitable in Sommerfeld's case, or as an inevitable foundation of any theory in Schrödinger's case.

Yet in both cases Bohr was aware of this deficient aspect of his account; he just did not think the criticism had enough weight in light of the account's heuristic value. He very well understood and appreciated the standpoint of the two physicists, but never allowed it to impede the inductive process that advanced the theory. He politely responded to Sommerfeld's concerns in 1924: "I should not like you to get the impression that my inclination to pursue obscure, and undoubtedly other false analogies makes me blind to that part of the formation of our conceptions that lies in *the unveiling of the systematic of the facts* [emphasis added]. Even if I were blind, I would only need to glance at your book to be healed" (Bohr to Sommerfeld 21 November 1924; Bohr 1972–2008, vol. 5, 1984, 38).

Intermediary constructive hypotheses are very close to the experimental context: bounded and limited by it, pulling it together with a more general master hypothesis. This was true of the CP when compared to the abstract model of the atom as the master hypothesis, the construction and the use of which was crucially mediated by this intermediary hypothesis. The model could, in various stages of its development, depart a great deal further from the experimental context and classical physical descriptions. Thus, the measure of the success of the CP was its empirical ability to use the key notion (i.e., quantum) figuring in the model to reach the lower hypotheses (i.e., using even the smallest details of spectral analysis experimentally reachable at the time, to paraphrase Bohr in Kragh 2012, 214). Bohr emphasized this as the most important aspect of the principle's adequacy (Kragh 2012, 214). Others understood this heuristic value of a mediating principle as well, as they used it or its results to advance their own accounts. Heisenberg summarized this nicely when explaining the nature of its use in relating his model of the atom with the most recent results of spectral analysis of the multiplets in the fine structure of spectral lines, long before he joined Bohr's Institute: "The fact that the entire model interpretation of a process can be calculated from purely empirical material is another brilliant achievement of the Bohr correspondence principle, which I am beginning to consider as important as the entire quantum theory" (Heisenberg to Landé, 29 October 1921, in Cassidy 1979, 211). In fact, the usefulness of the CP may have extended well beyond Bohr's model of the atom after the emergence of quantum mechanics (Kragh 2012, 217–19).

6: RECEPTION

The inductive circle was now closed. It started with a number of lower hypotheses induced from the experiments with spectroscopy and radiation, grasped via intermediary ones concerning Rutherford's model of the atom, atomic stability, Planck's law of radiation, and the correspondence principle that mediated them. It ended with the master hypothesis that refined and assimilated all these hypotheses into the model of the atom. The concepts figuring in the model were developed gradually from the bottom up—from atomic states or "terms" in producing spectral lines, ions, and electrons as manifestations of subatomic charged particles and quanta of radiation, to quantized orbits of the electrons and quantum numbers that more precisely characterized the atomic structure. Yet the result of this long inductive chain, and the impossibility of applying the standard mechanical and electrodynamical approaches it implied, have not been deemed very intuitive ever since. This is not surprising. Practicing experimentally minded inductivists of Bohr's kind have been aware for a long time that a novel hypothesis induced as the synthesis of broad novel evidence is bound to be perceived as counterintuitive.

Overall, the initial reactions were mixed. The complaints voiced by Rutherford (Rutherford in Bohr 1972–2008, vol. 2, 583) and Einstein (1917) were very similar in spirit to many comments made when Bohr introduced the principle of complementarity more than a decade later. Some asked an obvious question: How can electrons "know" where to jump once the atom absorbs energy, and which way should they move?[1] Many others found it "too bold and fantastic."

More precisely, the essential obstacle to wholeheartedly accepting a comprehensive model that provided such a precise agreement with the experiments was the need to divorce the mechanical laws of the electron's orbital motion from the radiation laws for which electrons were responsible. The precision came at the price of intuitiveness, as the predicted radiation frequencies of spectral lines were not represented by a suitable combination of mechanical motions: the electron's orbiting frequencies (Aaserud and Heilbron 2013, 173). The electron transitions resulting in radiation were not accountable mechanically. Bohr open the question of whether any sort of external causal process led to electron transitions across stationary states. He entertained the idea of acausal, inherently statistical process as well (Kragh 2012, 201), thus predating Max Born's similar treatment of Schrödinger's equation by more than a decade. Many other physicists, like Jeans and Henry Moseley, were instantly taken by the general applicability of Bohr's new model, as was David Hilbert. This was clearly its strongest suit (Pais 1991, 154).

When it came to accounting for the molecular structure, some saw the ever-increasing abandonment of the principles of classical mechanics as cowardly:

> For the complicated electrodynamics of electron rings revolving about a positive nucleus, Bohr had substituted the fiat that none should radiate; for the equally complicated electrodynamics of emission and absorption, he offered the simple incomprehensible formula. . . . Thomson felt this withdrawal as a betrayal. To him the invocation of the quantum was a fig leaf to cover ignorance of atomic structure and atomic processes. (Aaserud and Heilbron 2013, 191)

Yet what Thomson missed, and what some current critics are missing as well, is where exactly the power of the model lay: its precise agreement with a number of lower hypotheses, a case-by-case coherence, as it were, that no other model could match at the time.[2] The coherence of the model consisted in its ability to connect the parameters from the lower hypotheses, rather than in its reinforcement of old preferred general principles. The "renunciations" were tolerated in light of this, by physicists starting with Bohr himself, as mentioned. Those who focus on them now as a criticism of Bohr's approach to quantum phenomena are not really doing anything new; yet the criticisms are off the mark if they ignore the advantage the model had at the time, and what its explanatory power was based on.

To sum up, the sort of inconsistency that Bohr allowed for the sake of the heuristic value of the model was what Heisenberg, in conversation with Kuhn on this issue in 1963 (Heisenberg in Kuhn 1963), deemed "paradoxes"—high-level discrepancies, in Kuhn's parlance, in the theory that could not be avoided for the time being, though one might want to do so. In Bohr's atomic model, these were the classical and quantum aspects of the model that subsumed clusters of experimental results via intermediary hypotheses. New lower experimental laws exposed the limits of the principles, and an adequate master hypothesis had to reflect this for the time being. To Peter Debye's expectations that *a general principle* must connect quantum theory and regular electrodynamics, Bohr replied in the following way:

> I tried to say that the necessity of such a principle was perhaps not evident, that the problem which classical mechanics and electrodynamics had tried to solve perhaps was very different from the one which the phenomena confronted us with, that the possibility of a comprehensive picture should not be sought in the generality of the points of view, but rather in the strictest possible limitation of the applicability of the points of view (Bohr 1972–2008, vol. 2, 561).

Soon after the model was publicized and discussed, Walther Kossel (1914) argued that Moseley's experiments with X-rays in which he measured the wavelengths agreed with the predictions of the model. The spectral lines shifted across elements in accordance with the change of the number of electrons in the neutral atom. The splitting of spectral lines discovered by Stark was another confirmation, noted by Rutherford (Rozental 1968, 61) and later crucial for developing a generalized version of the CP. Rutherford also pointed out the relevance of the Zeeman effect as a new challenge to the model. Accordingly, Bohr set out to revise his model by performing an experiment with Hansen (figure 6). He then performed an experiment to see how his model could account for a new discovery by James Franck and Gustav Hertz (1913) that proved pivotal in the later development of quantum mechanics. Franck and Hertz discovered that mercury absorbs energy in quanta when bombarded with electrons, but they interpreted this fact as ionization because the atoms did not release energy below a critical value. Later, in 1917, it was discovered that ionization was a secondary process, and that Bohr's model accounted for the phenomenon (Franck and Hertz 1919).

The process of refining the model, with its refinements feeding on incoming new experimental results, some of which were initiated and seen through by Bohr himself, had started (Aaserud and Heilbron 2013, 167–70). In 1915 Bohr published a refined version of his model that accounted for new lower-level experimental hypotheses similar to those of Franck and Hertz. In addition, the last two papers of his trilogy were dedicated to the stable molecules and atoms with multiple electrons, a major concern at the time (Heilbron and Kuhn 1969, 213). This aspect of Bohr's work shows how he was adjusting the properties of his model to the experimental values—that is, the values of hydrogen molecule resonance frequencies

Figure 6. Niels Bohr, James Franck, and Hans Marius Hansen in 1921.
Niels Bohr Archive, Copenhagen

(ibid. 251) and the use of a fiddle factor to adjust it to the photo-effect (Aaserud and Heilbron 2013, 147)—and how he used the experimentalists' help to advance his model, as he used Hevesy's help to align it with the periodic table (Heilbron and Kuhn 1969, 253). Similarly, Sommerfeld and his group turned to Bohr's model and its refinements in their study of spectral lines, and devised a more realistic account of the hydrogen atom based on elliptical orbits (Seth 2010, 143, 163). In fact, Sommerfeld abandoned his method of problem solving and fully embraced modeling the atom in Bohr's style (ibid., 165) until 1919, when he turned to yet another approach of inferring theoretical propositions directly from the patterns of spectral lines without any intermediary hypothesis such as the CP.

Bohr's model of the atom was a master hypothesis, the result of a comprehensive synthesis of experimental results constructed through trial and error. It was thus inevitably provisional like any such master hypothesis, and new experiments were needed to refine it. As Alfred Landé noted, Bohr "was very dissatisfied with this model. I think he always had the idea that it was makeshift and something provisional" (1962). Heisenberg said, "For the first time I understood that Bohr's view of his theory was much more sceptical than that of many other physicists . . . at the time" (Heisenberg

1968, 94). Kragh acknowledged this view, stating that Bohr "conceived the model as preliminary and immediately began developing and modifying it" (Kragh 2013, 1). The master hypothesis was in fact a work in progress, continuously set against the experimental particulars, and the CP played a central role in this process.

The initial model of the hydrogen atom with one electron was introduced in 1913. Following Alfred Fowler's objection that the values of the wavelength for Bohr's model of the hydrogen atom did not agree with the experimental results (Fowler 1914), Bohr introduced a finite mass of the nucleus to exactly match the relevant results. Later, in the 1920s, he worked out models of heavy elements (Bohr 1972–2008, vol. 2 and vol. 4; Kragh 2012), multi-orbit models (ibid.), and elliptical models (Kragh 1979). The atomic model of the periodic system was thus achieved on the basis of the latest results in spectroscopy and chemistry.[3] Bohr finally moved on to a nonvisualizable quantum atomic model. In the early 1920s it was an eclectic hypothesis, a result of a useful inductive coordination of incoming empirical results and theory.

In all these stages of development, the model was crafted from the experimental results, via the correspondence principle. It was not developed through elaborate calculations.

> It is interesting to recollect how many physicists abroad thought, at the time of the appearance of Bohr's theory of the periodic system, that it was extensively supported by unpublished calculations which dealt in detail with the structure of the individual atoms, whereas the truth was, in fact, that Bohr had created and elaborated with a divine glance a synthesis between results of a spectroscopical nature and of a chemical nature (Kramers 1935, 90; translated by Kragh 2012, 300).

The resulting synthesis might or might not have been due to a "divine glance," but it certainly was the result of the substantially better than average skills of a laboratory and a scientific community leader who synthesized the overall experimental context into a master hypothesis.

Heisenberg had a similar impression of Bohr's method of developing his hypotheses, saying that "his insight into the structure of the theory was not a result of a mathematical analysis of the basic assumptions, but rather of an intense occupation with the actual phenomena, such that it was possible for him to sense the relationship intuitively rather than derive formally" (Heisenberg 1968, 95). And H. B. G. Casimir, the discoverer of the famous Casimir effect, said:

While pondering the philosophical problem of the description of nature, [Bohr] perfected to an ever higher degree the art of obtaining qualitative or semi-quantitative results without detailed calculations. This type of analysis that was partly based on an amazing skill in separating effects according to orders of magnitude was characteristic of all his work. In this respect he was much closer to experimental physics than more formal theoreticians (quoted in Rozental 1968, 110).[4]

Formal derivations and deep mathematical analysis were the next stage of development in the 1920s, starting with the provisional foundations that Bohr established and in which Dirac, Heisenberg, and others excelled.

7: THE SCIENTIFIC MODERATOR

The emergence of Bohr's model, and the relief it brought to the community as a useful tool of the analysis across phenomena, quickly changed attitudes to Bohr, and he soon established his central mediating role:

> Bohr, who had earlier met with considerable criticism and lack of understanding, had at this time become one to whom all listened with reverence, so that the discussions about the lectures were rather concerned with whether Bohr had meant this or that, than the matter itself. . . . During the decade 1913–1923, Bohr had undoubtedly been the leading scientist with respect to the whole complex of problems comprised by the new atomic theory which was erected on the quantum postulate (in spite of important contributions from other physicists) (Klein 1968, 84–85).

Once Bohr moved back to Copenhagen, with this newly acquired prominence in the physics world, he started lobbying for the funds he needed to start his own laboratory, of the sort that Thomson and Rutherford ran. His Institute for Theoretical Physics, founded in 1921, was a traditional physics laboratory. Bohr's official request for funding contained an explicit appeal for a new focus on the second inductive stage at the institute: in effect, experimentation in the service of a refinement and extension of, as well as a challenge to, the established model and its various aspects.

> Theoretical physics therefore now faces a task which can be justly characterized as the opposite of that which one had thought until a short time ago, namely to infer from the [experimental] information gained on the internal structure of matter the general laws. . . . Therefore it is necessary that the practitioners . . . carry out and guide scientific experiments in direct connection with the theoretical investigation (Bohr 1922c, 169).

Bohr established the institute as the symbiosis of theoretical and experimental activity revolving around the issue of the atomic structure. His

style obviously matched that of Rutherford and other prominent laboratory leaders, and could be characterized in the following way: "Although
there was no doubt as to who was the boss, everybody said what he liked
without constraint. . . . He was always full of fire and infectious enthusiasm
when describing work into which he had put his heart and always generous
in his acknowledgement of the work of others" (Andrade in Rutherford
1962). Bohr's role as a mediator and communicator of ideas, someone
who gathered the community, was an essential part of his character early
on, as was his interest in a hands-on approach to natural phenomena
(Rozental 1968, 16).

The critics, especially recent ones, often paint a picture of Bohr as an
established, grim, middle-aged, esoteric philosopher deep in thought, who
imposed his vision on others. Yet when Bohr built his model he was a
young, intellectually vigorous, and daring thinker with a vision considered
as wild as it was successful in its applications by many physicists. One
such physicist was Einstein, who praised Bohr's "unique instinct and tact"
in doing physics (Einstein 1949, 47). In a letter to Bohr dated 2 May 1920
he nicely summarized his favorable impression: "I understand now why
Ehrenfest is so fond of you. I am now studying your great papers, and in
doing so—when I get stuck somewhere—I have the pleasure of seeing your
youthful face before me, smiling and explaining. I have learned much from
you, especially how you approach scientific matter emotionally" (Einstein
in Kragh 2012, 203).

The mediating role of the young physicist was not only social; it was also
administrative. Bohr knew what he was doing in this respect too, and he
built his institute as a place for international physicists, handling much of
the administrative work himself. Heisenberg noted that "the administra-
tion of the Institute even then [in the mid-1920s] rested heavily on Bohr"
(Heisenberg 1968, 96).

Ultimately, Bohr was a central mediator between experimental and
theoretical work in his institute, and in the larger quantum physics community, until the 1930s. The kind of physics he did and his breakthroughs
owed a great deal to how he played that role, with his detailed understanding of both aspects:

From the very beginning Bohr had a complete understanding of the
stimulating reciprocal interaction between theory and experiment, and
a short time after his return to Denmark he got the idea of building the
institution where both theoreticians and experimenters could work in the
closest contact with each other. The result of this intimate collaboration,

which has continued through the years . . . surpassed all expectations and came perhaps as a surprise to those who had not gone into the question as thoroughly as Bohr did (Rozental 1968, 150–51).

Not surprisingly, then, given the interests of its director, the institute became a place of experimental and theoretical assessments and reassessments of quantum phenomena, with each result and every experiment directly motivating the next theoretical development, and vice versa. Every detail of fresh experimental results was considered, and every detail of a theoretical suggestion was immediately tried in light of new experiments (Frisch 1968). The experimental work at the institute spanned decades. The newest equipment and techniques were introduced as quickly as they were designed: Geiger counters, Enrico Fermi's use of beryllium powder, the use of radio phosphorus and radium, spectroscopic analysis, Van de Graff apparatus, an isotope separator, one of the first cyclotrons, later experiments with biological phenomena of interest, and especially the apparatus needed for the breakthrough experiments with fission.

The experiments were designed, often by Bohr himself, to test the challenges to Bohr's model of the atom and related ideas or to probe new phenomena that led, for instance, to the discovery of hafnium or fission. Bohr, too often seen as an abstract thinker clothing the physical theory in his favorite philosophical garment, was actually at the center of this experimental work, and he used it to refine and develop his two master hypotheses throughout his career. One of the later famous breakthroughs was the so-called compound nucleus: Bohr realized that the capture of a slow neutron by the nucleus was an indication of the nucleus's compound structure. The experimenters at the institute successfully tested this hypothesis by making use of various elements, such as gold (taken from Bohr's Nobel Prize medals), cadmium, and boron.

Bohr's urging for the proximity of and daily communication between mathematicians and physicists was related to the second stage of developing a comprehensive theory that was bound to be much more math-driven at some point. He managed to get funding to build a Mathematical Institute next to his Institute for Theoretical Physics: "This was a natural development of the idea that such disciplines as physics and mathematics, which could work hand in hand and be of mutual benefit, should be situated close enough to one another to ensure daily contact" (Rozental 1968, 150–51). Yet physics as a continuous almost daily interchange between theory and experiment in a tight community of theoreticians and experimenters, a community that gradually and effectively built an inductive structure of hypotheses, was a vision that could be vigorously pursued only until the 1930s in the field of fundamental physics. Since then, the requirements of laboratories for the study of elementary particles have become larger and more complex, while the breakthroughs in theory have increasingly relied on deep, abstract, and specialized mathematical work that needs long and arduous development on its own. Changes in the nature and time scale of interactions between experiment and theory continued throughout the second half of the twentieth century and into the first decades of the twenty-first.[1]

Bohr's communications with other scientists are particularly interesting, and they shed light on his role as mediator. He turned to Franck, Hevesy, Hansen, and Darwin for the latest insights on the experimental context, and he very closely followed all aspects of the experiments performed at the time, including those by Compton, Geiger, Wilson, and Carl Ramsauer. In the 1920s he corresponded with Schrödinger and eventually met him, and he worked closely with Heisenberg at the institute. His communications with these other scientists were not typical, however; they consisted of reflections on the issues and the voicing of concerns, followed by a retreat into a somewhat better informed standpoint on the nature of microphysical phenomena. This differed, for instance, from the 1920s correspondence between Einstein and Schrödinger, who agreed to disagree in their accounts of quantum states. The only thing they agreed on was that Bohr's emerging account of complementarity was wholly unacceptable. But Bohr entered into his communications with others with the clear aim of building a general theory that could absorb both the experimental insights and the essentially limited viewpoints accounting for them that others had articulated. From the very beginning, Bohr understood his role as a moderator and a mediator between the experimental context and conceptual insights. Fairly quickly, right after his work on the model of the atom was published, he "had become the principal consolidator of one of the greatest developments in the history of science" (Pais 1968, 219).

One side of the mediator's role was to repeatedly warn that we should not prematurely explain away discrepancies in two substantially different experimental contexts by postulating entities and their properties clearly prompted by just one of them. Furthermore, the reason for pronouncing a general hypothesis should not be strictly or primarily metaphysical. In Rosenfeld's words, the other side of Bohr's role was to "never try to outline any finished picture, but . . . patiently go through all the phases of the development of a problem, starting from some apparent paradox and gradually leading to its elucidation" (Rozental 1968, 117).

*

Thus, adequate master hypotheses were constructed from the experimental context: a set of varied experimental setups of relevant physical phenomena. But how exactly were particular existing hypotheses sorted out by the moderator from this complex web of hypotheses?

The so-called eliminative induction focuses on the binary process of selection (dilemmas), while the master hypothesis is induced from multiple intermediary and limited hypotheses. The key to understanding the eliminative inductive process is in understanding how to make the appropriate choice among rival or alternative suggestions. Rival accounts (e.g., the systems of Copernicus and Ptolemy) are taken to contradict each other, while alternative accounts (e.g., the systems of Copernicus and Kepler), though not contradictory, offer other possible explanations. Following this line of argumentation, Weinert (2001), Worall (2000), Laymon (1994), and Earman (1992) have attempted to explain why a scientist will choose one hypothesis over another and what kind of criteria guide her choice. Briefly stated, various empirical and some nonempirical constraints determine such choices. Each choice can be represented as a dilemma between two rival hypotheses,[2] one of which is dismissed in the process of deliberation.

Several case studies purport to demonstrate the use of eliminative induction in the history of science. For example, Weinert (2001) analyses twentieth-century theories of atoms, arguing that Rutherford and Bohr arrived at their respective models via such a method. He argues, for instance, that this led to Bohr's dismissal of Thomson's model and his incorporation of Rutherford's instead.

The choice between two hypotheses might be part of the inductive process in certain but not necessarily typical instances, and at a very abstract level of discussion about the subject matter, once a hypothesis has a clear rival or an alternative. For instance, in the development of quantum mechanics, as we will see, dilemmas of this sort emerged only in the late 1920s, once the provisional inductive construction and the foundations of the theory were built on the comprehensive experimental context. As far as the choice between Thomson's and Rutherford's models is concerned, the clear-cut choice that even Heilbron and Kuhn (1969, 212) entertained as a possibility never really occurred. Rather, as we have seen, as Bohr was in Rutherford's lab, he was exposed to Rutherford's unpopular model and could compare it with that of Thomson. He did so, and as a result he incorporated various aspects of both into his own emerging model—namely, those aspects that best agreed with the relevant lower hypotheses. Thomson's model introduced localized electrons, thus reinforcing the view of their existence along with Rutherford's model, and it postulated mechanical stability. Moreover, as noted previously, Bohr effectively transferred "Thomson's techniques to the nuclear atom" (Helibron 2013, 26). And finally, Thomson revised his model in 1910, turning it into a "doublet model," a peculiar mix of a dipole value and an electron in a circular orbit, and this influenced Bohr's work on Planck's law and magnetism (Heilbron and Kuhn 1969, 226–27).

As far as the elicitation of hypotheses from the actual experimental context goes, the following point seems to be overlooked in this general approach to induction: a scientist who is particularly good at utilizing the inductive method simultaneously keeps her eye on a variety of the experimental results she deems relevant, and a certain number of limited intermediary hypotheses elicited from them, in order to synthesize them all. This process does not seem to be adequately described as a sequential, step-by-step procedure wherein the physicist decides between pairs of rival or alternative hypotheses just to move on to the next dilemma. Rather, the physicist does it comprehensively, albeit gradually, crafting the master hypothesis so that it agrees with multiple experimental considerations which are not considered in isolation; think of the synthesis of Planck's law and the spectral analysis in Bohr's model. And the process is selective; it pays attention to particular aspects of the experimental context that she judges ought to be put together from various experimental setups and assimilated into the developing provisional hypothesis. This is apparent both in the development of early quantum theory and in the later development of quantum mechanics. This is precisely the kind of inductive trait in which Bohr excelled, and it enabled his mediating role in the development of the theory.[3]

In fact, the key decisions that result from clear-cut dilemmas could be detrimental, as the physicist will be forced to stick to a ready-made hypothesis rather than focusing on formulating a novel, refined one based on a comprehensive grasp of the experimental context. This is exactly the kind of danger that Bohr's approach managed to avoid while he synthesized his master hypothesis from various intermediary ones. He never fully opted for Rutherford's model, but incorporated certain of its aspects into his own model. The methodological "trick" at that stage of the inductive process is to avoid being stuck with a hypothesis that favors one of the experimental setups at the expense of another. We seek in vain for such clear choices between alternative hypotheses in Bohr's work. Thus, it may not be all that rewarding to approach Bohr's inductive method with such expectations; we might end up pointing to the failings of his conclusions, much like the harsh critics I mentioned previously.[4] Bohr's inductive method enabled him to juggle multiple intermediary hypotheses which, in turn, allowed him to build a comprehensive model as a master hypothesis based on a comprehensive experimental foundation.[5]

Similarly, a simplistic psychological account of how exactly Bohr came up with the model is offered by an otherwise exceedingly erudite historian of Bohr's work, John L. Heilbron (2013). Heilbron's analysis appeals to more tenuous factors and psychological analogies (e.g., quantum jumps of creativity and mind) that, he purports, describe the creative process that generated the model. A questionable assumption here is that there is an easy answer to the question of how Bohr got the crucial details on the relationship between frequencies and electron orbits (Heilbron 2013, 35). This is somewhat puzzling methodologically: even a psychological explanation should not be independent of the historical analysis of the process Heilbron himself offers. The historical context is, in fact, essential in such an explanation. And the context is such that Bohr favored a bottom-up, experiment-driven inductive process. Other external factors and influences Heilbron (2013, 53) tries to identify are possible extensions of this process, and certainly must have worked within its confines. Looking for them as the solution to the creativity puzzle is misleading. Such influences could have helped refine and accelerate particular aspects of the process in an indirect way, but they could hardly have provided a foundation for the physics Bohr developed. For example, only a fairly simplistic account that sidelines historical analysis could portray Bohr as a metaphysically driven physicist, as many philosophers do.

This is especially apparent in interpretations that posit the direct influence of a thinker on Bohr's method. For instance, Suman Seth offers paragraphs from Høffding's work to demonstrate similarity with Bohr's alleged commitment to discontinuity as a basic property of human thought that was translated into a guiding idea in constructing CP (2010, 99). Yet any other analysis of the vaguely related ideas of any other thinkers will be as informative—those of Kant, of C. I. Lewis, or of Talmudic thought spring to mind. Høffding must certainly have influenced Bohr and helped him achieve what he did by instilling in him a sense of philosophical inquiry: both a skeptical attitude and a continuous search for adequate ideas. But a result of this was that Bohr was taken by various ideas of various other philosophers, not just those of Høffding. If we could precisely identify a direct and substantial influence of a sole philosophical work or idea on Bohr's construction of the atom (and I don't think we can), this would be essential evidence for the most accomplished instance of application of a philosophical theory ever.

Metaphysical articulation was always post hoc in Bohr's work, guided by the inductive process. Bohr claimed that his work was at a safe distance from any one philosophical doctrine,[6] and this was demonstrably true. The instability of the atom and the discontinuity introduced in the model of the atom were certainly reminiscent of Høffding's philosophical ideas on the role of discontinuity in knowledge. Bohr could have drawn on these particular ideas for his final layer of motivation and could have used them to make his model more appealing to other educated minds. Yet, as we will see soon, when he devised his complementarity principle, he turned to a different set of philosophical ideas, a double-aspect theory pursued by Mach and by other philosophers with similar ideas like Spinoza and Leibniz, when those ideas were best suited to articulate the final principle. He did not stick to a particular set of metaphysical ideas, but he did stick to his inductive method. He sought the former to frame the product of the latter to make it more appealing to philosophically informed physicists—a majority at the time—and to a wider academic circle.

PART 3

Toward Quantum Mechanics

The Heisenberg-Bohr tranquilizing philosophy—or religion?—is so delicately contrived that, for the time being, it provides a gentle pillow for the true believer from which he cannot very easily be aroused.

—Albert Einstein to Erwin Schrödinger, 31 May 1928[1]

Interpretations . . . being gathered here and there from very various and widely dispersed facts, cannot suddenly strike the understanding . . . [and] cannot help seeming hard and incongruous, almost like mysteries of faith.

—Francis Bacon (2000, 38, xxviiiv)

8: QUANTUM CORPUSCLES, QUANTUM WAVES, AND THE EXPERIMENTS

This part of the book articulates the nature of Bohr's approach to quantum phenomena during the emergence of quantum mechanics, in the middle to end of the 1920s, as a synthesis of the theoretical approaches developed by others, especially Heisenberg and Schrödinger. Each approach was shaped by the specific experimental context in which it was developed. And it was against this experimental backdrop that Schrödinger's aim to present a general hypothesis based on his wave mechanics failed to materialize, following a debate with Bohr. Bohr conceived his complementarity account as a new master hypothesis against this backdrop as well (chapter 8). Moreover, each attempt to provide a link or demonstrate equivalence between these accounts, such as Heisenberg's uncertainty principle (chapter 9) or Schrödinger's proof of formal equivalence (chapter 11), was strongly subject to specific limits determined by the relevant experimental context. Heisenberg's uncertainty principle was articulated with Bohr's help as a constructive intermediate hypothesis primarily outlining the limitations of the supporting hypothesis—in contrast to the correspondence principle that served as "glue" between the experiments and the master hypotheses (chapter 9). Thus, Bohr's complementarity account was the result of an inductive-hypothetical process that can be subsumed under a few heuristic rules (chapter 10), but it was also a theoretical guide to the discovery of the quantum tunneling effect and remains such in current experimental work on the phenomenon (chapter 12).

Bohr's contribution to old quantum theory was a flashy breakthrough and it displayed his talent for inductively grasping the entire body of experimental results in the form of novel operational hypotheses produced at different levels, which nobody else had foreseen. As we have seen, it surprisingly and usefully tied together seemingly unconnected theoretical and experimental hypotheses. This later contribution did not match the piercing effect of Bohr's earlier contributions, but it illuminated Bohr's role as a successful mediator of various highly contending yet formally well developed approaches in the field, by provisionally and usefully reconciling them. Although provisional, this synthesis exerted a long-term influence, becoming the template of Bohr's particular experimentally-minded approach to quantum phenomena.

Chapter 8 traces how Heisenberg and Schrödinger developed their respective general accounts of quantum phenomena, and shows how they were turned into supporting hypotheses over the course of debates with Bohr, focused mostly on the new round of light/matter scattering experiments. It discusses the crucial details of these experiments as a backdrop of Bohr's emerging new master hypothesis, the essentials of which were outlined in his 1928 *Nature* paper. In fact, as we will see at the end of this chapter, perhaps more than in any other of Bohr's publications aimed at a specialist audience, the structure of the arguments in this 1928 paper, along with his 1925 *Nature* paper, makes very apparent the role the experiments played in the emergence of the complementarity account of quantum phenomena as a novel master hypothesis, as much as they trace the main steps of the inductive-hypothetical process Bohr employed.[1]

*

Bohr was motivated by multiple new experiments to gradually abandon his model of the atom and instead develop his complementarity account in the mid-1920s. These experiments implied completely different yet equally compelling accounts of microphysical states, as each was in agreement with specific ontological and epistemic viewpoints preferred by the physicists who pursued these accounts. Bohr's perspective led him to pick a middle road in deciphering the experimental particulars and the data by making sure he understood the "pull" of the opposed hypotheses. Meanwhile, he refrained from treating them as general if relevant experimental particulars elicited justified doubt. The complementarity account was largely a result of his repeated efforts to prevent hasty generalizations that were precluded by the experimental context.

The basic structure of the process that led to Bohr's development of complementarity is as follows. The group of experiments with light (wave) interference, on the one hand, and the scattering experiments establishing the photon-electron pairing (Geiger and Walther Bothe's work) at particular angles (Compton and Alfred W. Simon's work), on the other, led to basic but contradictory lower experimental hypotheses: the "paradox" in Bohr's (and Heisenberg's) terminology. More specifically, the former were instrumental in generating the wave-interference hypothesis, while the latter crafted a surprising (at least at the time) hypothesis on the conservation of energy and momentum in individual photon-electron interactions. Both were actually timid and limited experimental hypotheses, but they led to more substantial work. Wave mechanics modifying the notions of wave interference resulted from the former, while the quantum-corpuscular hypothesis, as well a substantially more elaborate matrix mechanics, stemmed from the latter (it also was closely tied to the discrete aspects of microphysical phenomena demonstrated by the lower hypotheses of spectral analysis). Heisenberg's uncertainty principle defined the limits and thus the applicability of these two supporting hypotheses with respect to the phenomena accounted for by the lower hypotheses. In the final step, Bohr applied Heisenberg's intermediary constructive and versatile hypothesis to synthesize the key aspects of the two opposed (paradoxical, in Bohr's and Heisenberg's parlance) hypotheses into his master hypothesis of the complementarity principle.

*

Heisenberg and Schrödinger took substantially different routes to their respective theories before Bohr assimilated them into his complementrity principle. Both theories were only gradually established, connected to various degrees to real experiments; and then tried in various toy models. Moreover, once the theories emerged it was not clear whether there was an inherent relationship between their accounts or between the formalisms on which they relied.

The approach to quantum states Heisenberg devised in 1925 begins with the particle-like properties as basic, actuated by the existence of the spectral intensity lines, but only in a very pragmatic sense. As for the epistemological constraints, unlike Schrödinger, Heisenberg did not think the microphysical entities should necessarily be visualizable in space and time. From the start, he abolished the bottom-line commitment to spatial continuity in explaining and understanding quantum phenomena. He also abandoned the commitment to the individuation of properties that characterizes classical particles in his approach to microphysical states.[2] His approach in constructing his account emphasized discrete properties of the observed phenomena; among these, the occurrence of discrete spectral lines of different intensities was central. Heisenberg was keen on devising an operational mathematical model without focusing on its "intuitiveness" of the resulting model. It was noted early that his theory could "be considered as a kind of phenomenological theory, as it poses for itself the task of establishing relations only between quantities that are observable in principle" (Thirring 1928, 385).

By introducing the quantized angular momentum of the electron, Bohr's model of the atom accurately predicted the spectral Balmer lines, as these corresponded to the rotational frequencies of the electron allowed by his model of the atom. Now, Heisenberg (1925) took discrete values of the spectral lines as the starting point—the primitives of his account, as it were—and gradually developed matrices to numerically account for them.[3] He initially derived a noncommuting law of momentum and position as discrete values, and soon after, with the help of Born and Pascual Jordan (Born, Heisenberg, and Jordan 1926),[4] he developed a matrix mechanics that accounted for these values and their noncommutative nature.[5] This was the major breakthrough in the emergence of the new quantum mechanics, and a decisive step in abandoning the old quantum theory.

It was not clear at the time what it all meant conceptually, but the new mechanics was quickly discovered to be closely related to various more detailed aspects of the experimental context. In particular, Schrödinger's emerging wave mechanics was not capable of accounting for Balmer lines as straightforwardly as matrix mechanics (Schrödinger 1926b, 30). Schrödinger thought this was merely a technical advantage (Schrödinger 1926a, 57), and this gave rise to his equivalence proof, to which we will turn later.

Heisenberg's commitment to consider only observed states was probably a commitment that followed rather than motivated his working out of matrix mechanics (Camilleri 2009). Yet his endeavor was somewhat instrumentalist in spirit, especially when we compare it to the nature of Schrödinger's overall project and the motivation behind it. It was also in accord with, if not directly inspired by, the philosophical views of logical positivists (Wolff 2014). Camilleri (2009, 53–54) remarks, that "Undoubtedly, there was a strongly positivist-empiricist element in Heisenberg's philosophy in the 1920s." To this he adds, "Heisenberg concluded his 1927 paper on a typically positivistic note." In any case, it was another distinct and rather boldly pursued methodological approach to the puzzling quantum phenomena, also cautioned by Bohr's general attitude. After the late 1920s, Heisenberg substantially moved away from this epistemological attitude, adopting instead a stance in line with Kantian philosophy (Camilleri 2009, part 3).

Clarifying his preferences, Heisenberg recalled in the 1960s, "I found in the formulae, which were the result of my collaboration with Kramer, a mathematics which in a certain sense worked automatically independently of all physical models" (Heisenberg 1968, 98). In contrast, Bohr "feared that the formal mathematical structure would obscure the physical core of the problem, and in any case, he was convinced that a complete physical explanation should absolutely precede the mathematical formulation" (ibid.). For a mathematical modeler like Heisenberg, the fact that the discrepancy between the orbital frequency of the electron and the frequency of the emitted radiation in Bohr's model could not be resolved by an adequate mathematical model was deeply unsatisfying. Such a mathematical solution could and did renounce orbits and individual classical states, and embrace quantum discontinuities as basic. This was not a commonly shared view, however. Kramer, for instance, resisted it, as Pauli's letter to Bohr indicates (Pauli 1979, 148). Pauli shared Heisenberg's attitude; they started from the experimental results in order to formulate a mathematical hypothesis capturing them, after Heisenberg tried to retain electron orbits of Bohr's atom as heuristic devices (Camilleri 2009, 23).

Not surprisingly, then, Bohr located Heisenberg's insistence on the in-principle failure to understand individual microphysical processes within the confines of newly emerging theoretical framework, as a confounding of Heisenberg's instrumentalist attitude stemming from it, not as a general outcome of analysis based on the entire experimental context. In a letter to Ralph H. Fowler in October 1926, Bohr is reluctant to embrace this strong antirealist stance in light of the new results, and clarifies the issue instead: the "final recognition of the impossibility of ascribing a physical reality to a single stationary state [is] a confounding of the means and aims of Heisenberg's theory" (Bohr 1972–2008, vol. 5, 15). He does not embrace Heisenberg's stance but instead puts it in the context of the pursuit of one feature of the complementarity hypothesis.

Heisenberg's methodological attitude was certainly not novel to Bohr. There were physicists who much earlier had insisted on simply organizing experimental results into a mathematical system—for example, the "energeticists" spearheaded by Mach who were criticized by Sommerfeld at the beginning of the century from a position similar to Bohr's (Seth 2010, 144). It was an epistemological attitude that Bohr, given his intellectual vigilance, must have thought out long before his debate with Heisenberg and Schrödinger. Moreover, along with Sommerfeld and Born, Heisenberg belonged to the group of physicists, labeled "virtuosi" in Einstein's parlance, who relied on calculation as a sufficient means to solve a problem.[6] Bohr's approach instead was driven by the goal of the all-encompassing master hypothesis, and was welcomed by many as such. In 1922, Ehrenfest openly celebrated Bohr's liberation of quantum physics from mindless calculation (Seth 2010, 186). And he indeed barely used calculation when revising and aligning his model with new experimental results.

Heisenberg (1925), however, famously took the data/theory dichotomy in a different direction, abandoning Bohr's gradual ascent from data to higher-level hypotheses. Instead, led by an instrumentalist attitude, he argued for building theory *exclusively* from observable quantities. Bohr's atom, as it was mediated by the CP, did not satisfy this demanding and very specific condition, nor did Bohr think it should. Heisenberg's attitude may have been a "*post facto* justification for the elimination of the electron orbit" (Camilleri 2009, 17) in the face of the mounting problems with Bohr's model, especially the increasing implausibility of the notion of electron orbits, though the "observability principle" was formulated already in the introduction to his 1925 paper. And Heisenberg's move could have been intended to "make his elimination of classical particle trajectories more agreeable to his contemporaries" (ibid., 18). But he did eliminate them, and although the initial motivation for this may have been similar to Bohr's reasons for taking experimental results as the foundation of building a new theory, Heisenberg did not dwell on the subtleties of Schrödinger's account and its merits, as Bohr did. Whether we call this attitude instrumentalism, a qualified instrumentalism, or pragmatism (e.g., Camilleri 2009) is less important than its obvious contrast to Bohr's and Schrödinger's respective approaches. These were very distinct "directions of inquiry," to paraphrase Serwer (1977, 245).

It is understandable, then, that Schrödinger complained (1926e, 46) about the unintuitive nature of Heisenberg's approach, saying it bypassed the principle of *Anschaulichkeit* (one's capacity of intuitively visualizing a spatiotemporal picture of physical states) of the continuity principle to which Schrödinger was committed. Yet if the particlelike properties were deemed mere appearances right at the beginning, instead of phenomena that could reflect a facet of the observed systems, it would be almost impossible to encourage development of the approach supported by matrix mechanics. This might be an undesirable outcome in the given context. This worry was reflected in Bohr's approach, but also in Pauli's. In a letter to Bohr, Pauli wrote that the alleged necessity of classical visualizations "should still never count in physics for the retention of a certain set of concepts. When the system of concepts is once clarified, then there will be a new visualization" (Pauli to Bohr, 12 December 1924, Pauli 1979, 186).[7]

What was at stake for Bohr was exactly how, not whether, physical reality could be ascribed to individual states. The conflation of Heisenberg's antirealism and primarily mathematically-driven instrumentalism with Bohr's pursuit of complementarity may be the main reason for the widespread understanding of Bohr as an anti-realist. There are certainly passages that may lead to this sort of understanding, if we take them outside the overall context of his approach and work, and the precision that its target audience required. But Bohr's insights did not cause him to abandon an interpretation of the nature of individual microphysical states. They led him, however, to abandon the principle of continuity as the exclusive ontological principle guiding such an interpretation—a key feature of Schrödinger's account to which we now turn, and the account that Bohr eventually assimilated as the second supporting hypothesis (the quantum-corpuscular being the first) grounding his complementarity.

In any case, in contrast to the conceptual problems of "space-time coordination" that underlined wave-mechanical hypothesis, Bohr saw matrix mechanics as "limited just to those problems, in which applying the quantum postulate the space-time description may largely be disregarded, and the question of observation in the proper sense therefore placed in the background" (Bohr 1928, 585). The individual classical states and classical trajectories were removed and replaced by quantum discontinuities as basic states in Heisenberg's theory. In short, Bohr gradually started thinking of such a theory as an intermediate supportive quantum-corpuscular hypothesis: an "immediate expression" of the experimental hypothesis of the conservation of energy, the view he crucially drew from the scattering experiments by Walther Bothe and Geiger and by Compton and Alfred W. Simon (ibid., 587). Heisenberg eventually concurred with this, stating in his letter to Pauli, for instance, that "the theory of light quanta and even the Geiger-Bothe experiment is essential" (Heisenberg to Pauli, Bohr 1972–2008, vol. 6, 17). We will look at this in more detail shortly but let us see first how Schrödinger took a rather different route to his own account.

*

In 1926, Schrödinger (1926a, 1926b, 1926c, 1926d) published his famous four papers that, among other contributions, established the wave equation as the central formalism of quantum mechanics. In this series of papers he also advanced his initial wave-mechanical interpretation of microphysical states.

Initially, as discussed earlier, in a stationary state of the atom postulated by Bohr's model, an electron behaves as a classical particle orbiting around the nucleus. Yet absorption or an emission of energy results in a discontinuous transition to a different orbit. Like many other physicists, Schrödinger found these "quantum jumps" unsatisfying—and offered an alternative framework of transitions of atomic states, so that the space-time continuity of the microphysical process during the alleged orbital transitions could be preserved. His main motivation for this attempt was, he said, that if continuity was not preserved by "relinquishing the ideas of 'position of the electrons' and 'path of the electron' [then] contradiction is so strongly felt that it has been doubted whether the phenomenon in the atom can be described in the space-time form of thought at all" (Schrödinger 1926b, 27). He added the following very strong formulation underlying his approach: "From the philosophical standpoint, I would consider such a definitive decision of this sort to be equivalent to complete surrender. For we cannot really alter our manner of thinking in space and time, and what we cannot comprehend within it we cannot understand at all. There are such things—but I do not believe that atomic structure is one of them" (ibid.).

This credo was a prominent feature of the four papers Schrödinger published in 1926. As we have seen, Poincaré endorsed this attitude during the formation of old quantum theory, but Planck's intuitions certainly went the same way. At the Solvay meeting, Planck characterized the introduction of discontinuities prompted by the photo-effect as inevitable, as "ruining the foundations" provided by the agreement of the interference experiments and Maxwell's theory (Aaserud and Heilbron 2013, 144). More than a decade later, Schrödinger pushed this sentiment as far as he possibly could at the time.

As Linda Wessels has made clear, Schrödinger was primarily committed "to finding a coherent description of microsystems" (Wessels 1979, 272). This was also his main goal in his early work, as Christian Joas and Shaul Katzir (2011) have convincingly argued.[8] He tried to explain theoretical phenomena, rather than merely describe or predict them (Joas and Katzir 2011, 51), and many of his accomplishments were of a speculative nature (ibid., 52). He seems to have adopted this approach from Boltzmann, who thought of the aim of physical theories as providing clear statistical atomistic explanations in the form of intuitive pictures (Joas and Katzir 2011, 44). Both Schrödinger and Boltzmann "believed that by ascribing reality to [these] hypotheses"—that is, sticking to particular intuitive principles and pictures—"and by following their consequences for other phenomena, they would be able to 'learn something new' about the hypothetical entities" (ibid.). Schrödinger turned this into a combination of a strong scientific realist slant—the attitude that physical theories should be understood as insights into real physical states, and that this ought to guide the theoretical pursuit—with a metaphysical, somewhat holistic understanding of physical reality in which human observer is inevitably immersed (Bitbol 1996, 13–14).

In any case, the approach was strongly anti-inductivist, with little affinity for any heuristic goals. Theoretical tools were not to be treated as heuristically driven, or at least had to quickly transcend their heuristic value. Schrödinger's insistence on the *Anschaulichkeit*, the intuitive grasp of any concept of the microphysical, was an explication of such a metaphysically driven devising of the hypotheses. From the very beginning of developing his program, he focused on the nature of physical explanations of quantum states (Joas and Lehner 2009, 344). This meant developing the account within the framework of continuity and Boltzmann's understanding of statistics. It was this commitment that led him to so stubbornly pursue the wave-mechanical interpretation of quantum phenomena, supported by the principle of spatiotemporal continuity (Schrödinger 1926a, 27). In his four papers, he characterized the basic properties of interacting atomic systems in terms of the so-called characteristic frequencies (E/h). Instead of abrupt transitions during absorption and emission of energy, he argued, the atom's activity is accounted for in terms of wave vibrations. As a result, he wrote, "it is hardly necessary to emphasize how much more congenial it would be to imagine that at a quantum transition the energy changes over from one vibration to another, than to think of a jumping electron" (Schrödinger 1926a, 10).

The difference between the energies of two atomic stationary states, explained by Bohr in terms of the electron's quantum "jumps" in orbit, results from the exchange of energy between two vibrating states, characterized by appropriate modes of vibration. This wave-mechanical setup enables one to describe two physical states as resultant frequencies: the normal modes of the systems' vibrations constitute the energy exchange and, accordingly, account for what Bohr's model characterizes as orbital transitions. Most importantly, such wave-mechanical processes do not seem to violate space-time continuity.

Given his deep philosophical motivation, Schrödinger boldly pushed his general account even further.[9] The classical mechanical method ascribes n particles to every point in q-space.[10] For Schrödinger, however, each alleged "particle" should be attributed a wave function. He clarified this by introducing the analogy of the failure of geometrical optics, where any attempt to trace the incoming ray of light in the neighborhood of the diffraction patch is meaningless (Schrödinger 1926b, 26). In dealing with very small wavelengths, he stated, the classical mechanical equations describing the underlying mechanics of the behavior of particles in the electromagnetic field become as awkward in accounting for the true nature of the microphysical world, as the optics of rays does in explaining the phenomena of diffraction. Hence, "We must treat the matter strictly on the wave theory, i.e. we must proceed from the wave equation and not from the fundamental equations of mechanics, in order to form a picture of the manifold of the possible processes" (ibid., 25). Thus, he used the wave theory of light and the failure of geometrical optics as analogical intermediary hypotheses to support his general account of microphysical states. Yet his spirited attempt at the master hypothesis had a different slant than that of Bohr: he was crucially led by the principle of continuity to formulate it as an alternative to Bohr's model.

To top it all off, consistent with Schrödinger's metaphysical framework, a whole range of "paths" stretch in all directions within the classical "path" of the $3n$ spatial continuum. In fact, there is no exact point of phase agreement to which we could refer, and it is this phase agreement between the waves of the group that determines the location of a particle in q-space. Thus, Schrödinger's conclusion that "we can never assert that the electron at a definite instant is to be found on any definite one of the quantum paths, specialized by the quantum conditions" (Schrödinger 1926b, 26) led him to suggest an explanation of quantum phenomena in terms of continuous wave interactions. This required invoking the manifold of particle paths, which can be examined by analyzing the properties of the wave function. Schrödinger triumphantly concluded: "All these assertions systematically contribute to relinquishing the ideas of 'place of the electron' and 'path of the electron' in q-space" (ibid., 26).

The particles in an ideal gas that were counted by Einstein-Bose statistics, and which others did not find disagreeable (just because particles were characterized by rather strange statistics) to such an extent as to seek alternatives, did not satisfy Schrödinger. Instead, he modeled particles as wave modes to account for the indistinguishability of particles postulated by Bose-Einstein statistics, thus preserving the continuity principle and, as Joas and Christoph Lehner point out (2009, 344), returning to Boltzmann statistics.

*

Bohr was taken by this new approach. but remained cautious. He invited Schrödinger to his institute to discuss the burning issues in a wider and more experimentally-minded context. The spirited meeting between them ultimately spelled the end of Schrödinger's pursuit of a wave-mechanical approach to microphysical phenomena as an attempted master hypothesis for at least several years, though the wave equation as a formal tool became a mainstay of the emerging quantum mechanics. In fact, the meeting was so spirited that it seemed to contribute to Schrödinger's sudden exhaustion and illness (Bohr 1972–2008, vol. 5, 10–11). But Bohr continued pursuing his argument with his bedridden guest.[11]

Given the forcefulness of Bohr's persona—which, unlike Schrödinger, those residing in and frequenting the institute took for leadership—Schrödinger was under the impression that Bohr's criticism of the wave-mechanical approach was wholesale and categorical. He even expressed deep regret that he ever got involved in the issues of atomic theory.[12] Yet Bohr's criticism was very specific and, as such, it fit into his new emerging master hypothesis. On 23 October 1926, after Schrödinger found that his impressions had settled down, he sent a letter to Bohr commenting on the result of the debate, admitting defeat and pointing out the reasons for it:

> It is possible that the stubbornness, with which in our dialogues I continued to adhere to my "wishes" for a physics of the future, in the end may have left you with the impression that the general and specific objections that you raised against my views had not made any real impression on me. That is certainly not the case. In a certain sense I can say the psychological effect of these objections—*in particular the numerous specific cases in which for the present my views apparently can hardly be reconciled with experience*—is probably even greater for me than for you (Bohr 1972–2008, vol. 6, 12; emphasis added).

Schrödinger conceded that Bohr's argument refuted his pursuit of wave mechanics as a general account of microphysical states, and that it did so in light of the "specific cases" that could "hardly be reconciled with experience"—that is, with the existing experimental results. Which specific cases he was thinking of and why Schrödinger's wave-mechanical approach could not be reconciled with them will be clarified in due course; but we get a hint from Heisenberg's letter to Pauli:

> Just as nice as Schrödinger is as a person, just as strange I find his physics. When you hear him, you believe yourself 26 years younger. In fact, Schrödinger throws overboard everything [that seems] "quantum theoretical": photoelectric effect, Franck collisions, Stern-Gerlach effect, etc. Then it is not difficult to make a theory. But it just does not agree with experience (Bohr 1972–2008, vol. 6, 10).

Max Born's suggestion was similar, making it apparent that this was a wider and more appealing argument Bohr had pursued in the debate with Schrödinger: "On this point I could not follow him [Schrödinger]. This was connected with the fact that my Institute and that of James Franck were housed in the same building of the Göttingen University. Every experiment by Franck and his assistants on electron collisions appeared to me as a new proof of the corpuscular nature of the electron" (Born 1961, 103).

Thus, the "numerous specific cases" included, at the very least, Franck-Hertz and Stern-Gerlach's experiments and the photoelectric effect. Bohr also took into account Ramsauer's (1921) experiments with the atoms of gases. Yet the new series of Compton's experiments—after the first famous breakthrough series, the results of which had just become available—must have delivered the decisive blow. Thus, the final abandoning of the notion of *physical continuity of microphysical processes*, to which most physicists working on quantum theory at the time were in fact committed one way or another, was prompted by this second series of light-scattering experiments.[13] Schrödinger (1926a, 27) insisted on the notion of continuity as the bottom-line principle in understanding and even formalizing microphysical processes. But it was also understood as a generally applicable working hypothesis by other physicists, with the exception of Einstein. Finally and crucially, it was the motivation behind Bohr's own precomplementarity attempt, which he pursued with Kramers and Slater (Bohr, Kramers, and Slater 1924a, 1924b), to preserve continuity at least as a statistical trait of atomic interactions.

These experiments were performed by Bothe and Geiger and by Compton and Simon independently and with different experimental techniques. As a result, Bohr reduced the continuity-based approach to microphysical phenomena to an intermediary hypothesis with limited use within the overall experimental context, and an emerging master hypothesis of complementarity. But before we look at the details of these crucial and fascinating scattering experiments which pioneered the experimental technique of cloud chamber, I will briefly review the wider context of the debate.

The first and now more famous set of Compton's experiments with scattering, preceding this decisive second set, suggested that the energy and frequency in interactions of light and matter should be treated as classical interactions. Compton (1922a, 1922b, 1923a, 1923b, 1923c, 1923d) induced the existence of the electron recoil in the scattering from the quantum-corpuscular assumption and the value of measured charges. He performed the experiments with scattering of light by electrons using X-rays, measuring scattering angles and discovering recoil electrons by measuring the outgoing charges. The momentum in the experiments turned out to be conserved, on average, and light seemed to scatter from the matter in the way a classical mechanical account would suggest. To put it simply: on average, photons and electrons scattered in the way we would expect classical particles analogous to billiard balls to scatter.

When it came to the expectations of the still experimentally unprobed individual interactions, however, the attitude was almost unanimous. Bohr, Schrödinger, Darwin, Compton, and others agreed that Einstein's quantum-corpuscular hypothesis, confirmed much earlier by Robert A. Millikan's (1916) experiment with the photo-effect, should be abandoned in the context of individual interactions. On average, light and matter (electrons) interact in accordance with Einstein's hypothesis, but this is very unlikely in individual cases. There were no lower experimental hypotheses that suggested otherwise. Accordingly, each physicist developed a different intermediary hypothesis as to how this abandoning could be achieved and continuity preserved in individual interactions. In effect, their resulting pet hypotheses were part of a diverse set of generally wave-theoretical approaches that aimed to accommodate the results of the first series of Compton's scattering experiments, while expecting the second series to confirm them in terms of individual interactions. The physicists offered accounts of either classical or probabilistic waves, or a combination of the two. It was expected that the experimental testing of individual micro-interactions, the next series of scattering experiments probing the individual processes, would be useful in confirming one of these theories. Not even Compton expected a serious challenge to the wave-theoretical approach, an outcome that would result in a dramatic change in their thinking about the foundations of quantum theory.

In any case, the experiments by Compton that preceded and inspired the second series left open the question of the precise nature of the individual processes that might resolve the particle/wave quandary. As Heisenberg stated, "Even such important discoveries as the Compton effect . . . only sharpened the difficulties and contradictions." Moreover, "the central point of discussions at the time was dispersion theory; the theory of scattering of light on atoms" (Heisenberg 1968, 97). This was vigorously discussed at the Institute for Theoretical Physics in Copenhagen. The anticipated second series of experiments was open to two opposing possibilities: (1) that the incident wave interfered with the electron wave, and that this would produce variations in angles of scattering; (2) that, as the quantum-corpuscular hypothesis pursued by Einstein since 1915 predicted, the particle scattering would be exhibited only at momentum-preserving specific angles of the scattered photon and recoil electron, analogous to those in the first series of experiments.

In this second series of experiments, one set was performed by Bothe and Geiger (1925), using Geiger counters (figure 7), and another by Compton and Simon (1925), using the cloud chamber (figure 8). Both experiments delivered stunning results. They were fairly quickly and almost unanimously understood as forceful evidence that the scattering in individual interactions was, in fact, in agreement with Einstein's prediction. And this sort of scattering was seen as demonstrating the impossibility of wavelike "communication" between atoms as posited by the Bohr-Kramers-Slater theory. Soon after Bohr learned the results of Compton and Simon, he said they indirectly but convincingly demonstrated that Schrödinger's insistence on the continuity of microphysical processes was implausible: the momentum in interactions between light and matter turned out to be particlelike.[14] Schrödinger, however, resisted this conclusion even after Compton and Simon's experiment (Perovic 2006, 288). Bohr was initially skeptical himself (Beller 1999, 121) about conclusions based on Bothe and Geiger experiments alone, as they were restricted to the ninety-degree incoming photon angle. The photons scattering at ninety degrees were detected with two Geiger counters located opposite each other, and their number tallied with the number of the incoming electrons. Compton and Simon used track measurements in Wilson's cloud chamber instead, in order to detect scattering of photoelectrons at angles other than ninety degrees.

The forcefulness of Schrödinger's attempted refutation and replacement of Bohr's model, stemming from his conviction in the intuitive and metaphysical bottom line of a physical theory, matches the strength of the impression Bohr's argument left on him. Yet Bohr was not unsympathetic to Schrödinger's intention to provide a view that ascribed reality to the states of the system in terms of wave mechanics. Nor was he alone in this. Sommerfeld had abandoned the modeling approach to physical phenomena around 1919, and had turned to drawing theory from data in a more direct way, in particular from spectral line data. Wien saw the hope of overcoming such an "irrational" search for musical-harmony-like patterns in the data precisely in Schrödinger's attempt "to ascribe whole numbers of the quantum theory to similar characteristic vibrations. . . . Here number mysticism would be supplanted by the cool logic of physical thought" (Wien 1926, 15). Bohr's reservations about the wavelike nature of atomic interactions were novel; he had strongly opposed Einstein's quantum-corpuscular account of radiation until 1925,[15] and the quantum-corpuscular treatment of the photo-effect had not become an integral part of any iterations of his atomic model.[16] Instead, Bohr was

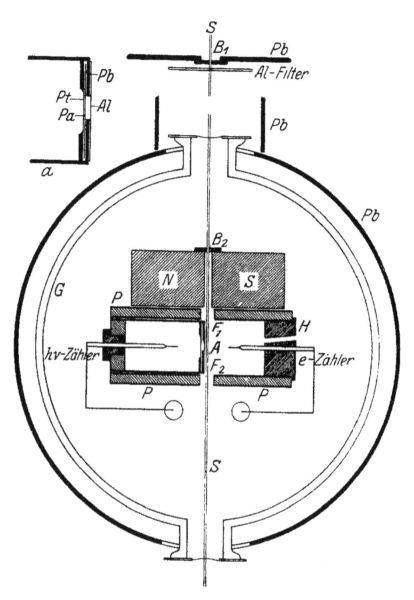

Figure 7. Bothe and Geiger performed a series of experiments with photon scattering. Photons scattered off the electrons at an angle of ninety degrees are detected with two Geiger counters located opposite each other. The number of photons is tallied with the number of the incoming electrons.

From W. Bothe and H. Geiger, "Über das Wesen des Comptoneffekts: Ein experimenteller Beitrag zur Theorie der Strahlung," *Zeitschrift für Physik* 32, no. 1: 639–63. Copyright 1925. Reprinted by permission of Springer Nature.

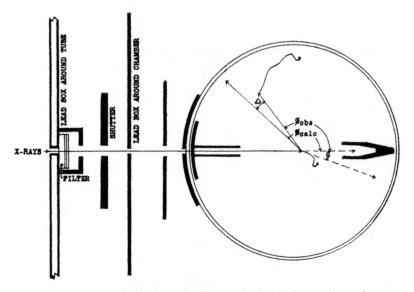

Figure 8. Compton and Simon used Wilson's cloud chamber to detect the scattering of photoelectrons at angles other than ninety degrees.
From Arthur H. Compton and Alfred W. Simon, "Directed quanta of scattered X-rays," *Physical Review* 26, no. 3 (1925): 289. Copyright 1925 by the American Physical Society. Reprinted with permission.

skeptical of accepting the wave-mechanical formalism and model as the exclusive basis for a general account of microphysical states, because of the "numerous specific cases." He did not refute the model wholesale, but he wanted to show its limitations. Bohr did not embrace Heisenberg's anti-realist confounding, though Schrödinger initially interpreted his skeptical attitude otherwise.

*

After these scattering experiments, Bohr started developing new master hypothesis. In his 1928 *Nature* paper, Bohr explicates the two key sets of experiments—those on the scattering of light and those on light interference—as the basic experimental particulars, explaining them as "facts" that together comprise the basis for second-stage inferences. We will discuss this milestone paper at length, but it suffices to say for now that Bohr begins, "In fact, all our knowledge concerning the internal properties of atoms is derived from experiments on their radiation or collision reactions" (Bohr 1928, 586). Meanwhile, "the interpretation of experimental facts [i.e., the second stage] ultimately depends on the abstractions of radiation in free space, and free material particles." Thus, the induced higher-level hypotheses that involve more abstract physical concepts not figuring in the lower-level experimental hypotheses are derived from the two sets of experiments. At the same time, the proper explication of the abstract terms' exact meaning (within the context of supporting and constructive intermediate hypotheses) is the job of a new master hypothesis, eventually labeled the complementarity principle.

The insistence on the bottom-up construction of the theory throughout the text is not simply a rhetorical tactic; rather, it represents the structure of Bohr's thought. This becomes apparent in his detailed characterization of the experimental grounds of the intermediary hypotheses that led to his model of the atom, its demise, and finally his new master hypothesis. Already in his 1925 paper, Bohr had explained why the choices of experimental particulars and cross-experimental comparisons at the level of lower experimental hypotheses were crucial for generating adequate intermediary hypotheses.[17]

After a vigorous conversation with Bohr, Schrödinger realized that his interpretation was doomed for the time being. The views of the physics community quickly converged on Bohr's conclusions, aligned with the direction of his emerging master hypothesis, later to be labeled the complementarity principle.

Returning to the wave hypothesis in Schrödinger's spirit before throwing the towel, however, we might continue to insist that discontinuities are solely appearances, even in these experiments. Thus, in general, the tracks within the cloud chamber could be still understood as determinate tracks left by a spherical wave. Nor is it necessary to immediately conclude that discrete segments of the wave, including those produced when a scattering takes place, make those tracks analogous to a particle whizzing through the cloud chamber, which would imply particlelike interactions of radiation and matter. According to an argument that must have contributed to Bohr's initial pause, what we actually observe in such a case could be a trace of the whole atom that ionizes consistently with the probability given by the wave equation. Following this line of argumentation, in the scattering experiments performed by Bothe and Geiger, the discovered scattering angles might be understood as *particlelike interference*, not the statistically distributed interference of wave fronts, though the latter was expected to be (but was found not to be) a mark of the interaction between continuous wave fronts (Kidd, Ardini, and Anton 1985). In fact, Mott (1929) developed such an alternate account of the results in the late 1920s, and Schrödinger followed his lead in the early 1930s. Bub (1974, 49) claims that

John von Neumann could have argued in favor of the same point, but did not because of the pressure of the Copenhagen interpreters. An alternative explanation is that von Neumann was well aware of the details of the argument Bohr developed in his conversation with Schrödinger.

Bohr refused this line of thought despite its theoretical subtlety, for the usual reasons of experimental comprehensiveness. Yet his argument was interpreted as intellectual bullying (Bub 1974; Beller 1999), but this was probably only due to misunderstanding of his inductive method. Bohr's arguments and approach were much closer to what Francis Bacon called "interpretations," which "by contrast [to anticipations] are gathered piece by piece from things which are quite various and widely scattered, and cannot suddenly strike the intellect" (Bacon 2000, xxviii, 28). Bohr's motivation, and the advantage of his inductive method over the one that built a metaphysically driven hypothesis, becomes apparent in this episode.

Bohr stated that the results of the Compton-Simon experiments with the cloud chamber demonstrated "the connection demanded by the light-quantum theory between the direction in which the effect of the scattered radiation is observed and the direction of the velocity of the recoil electrons accompanying the scattering" (Bohr 1925, 848). Schrödinger's account did not acknowledge this.[18] Following his experimental-context-driven inductive pursuit, however, Bohr also argued that "the suggestion [on the light-quantum approach] does not offer a satisfactory escape from the dilemma" between the wave approach and the light-quantum approach constantly discussed at the Institute for Theoretical Physics, precisely because of the insights provided by Schrödinger's hypothesis that Bohr treated as intermediary. On the one hand, in light of these experiments he gave up the account of probabilistic waves that he, Kramers, and Slater had developed; but on the other hand he "felt correctly that the apparent dualism was so central a phenomenon that he thought it should be the natural starting point for any interpretation" (Heisenberg 1968, 104).

Bohr explains (1925, 848) that scattering experiments with coincidence techniques demonstrated the "pairing" of electrons and photoelectrons, but he adds that, in conjunction with the experiments in the cloud chamber that probed the angles of scattered electrons, they effectively demonstrated the adequacy of the quantum-corpuscular account of the process. Thus, two sets of different observations in experiments with individual radiation scattering processes demonstrated limited experimental "facts." That is, the lower experimental hypotheses, namely the pairing and the scattering angles, jointly demonstrated that conservation laws obtain in interactions of light and matter and, accordingly, showed the adequacy of the quantum-corpuscular account (in the 1928 article this is treated as an intermediate hypothesis) as opposed to the statistical one. He reiterates this later in the 1928 paper (Bohr 1928, 584).

These summaries of the experimental context are a direct response to Schrödinger's ongoing effort to synthesize a wave-mechanical hypothesis from the experimental particulars in these two experiments. This required an experimental hypothesis based on the pairings and the probed angles in two sets of experiments, other than the one establishing conservation of energy and momentum in individual interactions. (This hope ended with Schrödinger's paper on the Compton effect, in which he admits a deficiency of his view in this respect.) The choice of the particulars and the way in which they were connected were crucial here, and that is why Bohr made sure to describe the way they were drawn at each level.

As we will see, the principle of complementarity was essentially a result of the synthesis of two of Bohr's viewpoints: the first embracing the light-quantum (supporting intermediary) hypothesis, and the second skeptical of its generality due to the quantum-corpuscular (supporting intermediary) hypothesis. The critics see the conjunction of these two viewpoints as decisive evidence that Bohr was a philosophical obscurantist who should not have blocked alternative approaches, such as those of Schrödinger or Heisenberg, by imposing his own wishy-washy syncretism. Yet such attitudes prevent the critics from understanding that Bohr's statements were a result of his inductive method. The method tamed two hasty generalizations, as it were, by fruitfully turning them into intermediary hypotheses, each of which connected certain sets of the lower hypotheses to the new master hypothesis. Bohr characterizes two intermediary hypothesis in heuristic terms when he says, "The means for a general consistent utilisation of the classical concepts in the quantum theory have been created through the transformation theory of Dirac and Jordan, by the aid of which Heisenberg has formulated his general uncertainty relation," adding that "in this theory also the Schrödinger wave equation has obtained an instructive application" (Bohr 1928, 587).

9: THE UNCERTAINTY PRINCIPLE
AS AN INTERMEDIARY HYPOTHESIS

There are numerous interpretations of the uncertainty principle, one of the central tenets of quantum mechanics.[1] Most are post hoc accounts concerned with the so-called foundational questions of quantum mechanics, but I am primarily interested in the formative process that led to it, and the role it played in the further development and finalizing of Bohr's complementarity principle following the formation of wave and matrix accounts of quantum phenomena.[2]

The uncertainty principle initially mediated between two rather general ones: wave and matrix hypotheses on the one hand, and the experimental phenomenon of tracks left in the Wilson cloud chamber by free electrons, whatever they might be, on the other, for which "neither matrix mechanics nor wave mechanics could as yet account" (Camilleri 2009, 46). Matrix mechanics was too abstract, and a mathematical model too pragmatic, to directly address the issue, while the wave-mechanical account was developed as an extension of the continuity principle. The apparently definite motion of particles was not so much at odds with these two quantum-mechanical hypotheses as it was unaccounted for by them.

On their own, the two accounts could not directly address the dilemma of the tracks in the cloud chamber—that is, the question of whether they were tracks of a particle that whizzed by, the trace of an ionizing cloud, or something else. The uncertainty principle (also labeled the indeterminacy principle) gradually bridged the two, starting as an analogy—as constructive intermediate hypotheses do—with the understanding of the notion of simultaneity first expressed in the special theory of relativity (Camilleri 2009, 94–95), developing into an operational principle (ibid., 86), and finally being amalgamated into Bohr's master hypothesis as a full-blown constructive hypothesis. The last stage of its development was the result of Bohr and Heisenberg's joint effort. During the debates at Lake Como, Heisenberg acknowledged this, saying, "The physical interpretation of the uncertainty relation . . . and its relationship with the general points of view raised by Bohr have been made entirely clear for the first time through the investigations of Bohr" (Heisenberg in Bohr 1972–2008, vol. 6, 141).

A large chunk of Bohr's 1928 paper is dedicated to exposition of the uncertainty principle (Bohr 1928, 582) as the expression of the "paradox." Unlike the CP, it played the role of a constructive intermediary hypothesis that clarifies by pointing out the exact nature of the limitations of both supporting hypotheses, rather than by producing ever more precise applications of supporting hypotheses, as the CP did. Bohr sees "the paradoxical character of the problem of the nature of light and of material particles" (Bohr 1928, 582) as unavoidable, even at the level of the individual microphysical, states in a protracted general treatment of the wave-particle duality.[3] Bohr's basic argument goes as follows. Wave packets emerging from a source one after the other, and consisting of sinusoidal waves, can be described by the wave number $\sigma = 1\text{cm} / \lambda$, which tells us how many crests are contained in a given wavelength interval. The wave packets of the same frequencies in the dispersion medium move at different speeds, thus virtually "reaching" each other. The superposition is a set of all such packages. The change in the position of such a "super-package" must be at least equal to the change in the wave number in the length interval: $\Delta x = 1 / \Delta \sigma x$ (y, z, t). The de Broglie wave is associated with the particle as the basic energy package. The basic particle action package is then $\lambda = h / p$ $(p = h / \lambda)$. So, for such a wave super-package, we have: $\Delta x \Delta \sigma \geq 1$. Our smallest unit in which the wave is packed is h; $(\Delta x \Delta \sigma \geq 1)$ times h gives $\Delta x (\Delta \sigma h) \geq h$, $\sigma = 1 / \lambda$, hence, $\Delta x \Delta p \geq h$, the basic expression of the uncertainty principle.

Bohr also develops what is now a standard presentation of Heisenberg's gamma-ray microscope thought experiment (1928, 582), and applies it to the Compton effect (1928, 583, 584) to emphasize the ambiguity of the scattering of light off matter processes, even though the experimental results "find suitable expression" in the quantum-corpuscular intermediate supporting hypothesis. Heisenberg introduced the thought experiment in the operational phase of the development of the principle to illustrate the limits of precision in defining the basic parameters of position and momentum in quantum mechanics. In the experiment, the instrument detects an incoming photon bouncing off an electron. This detection is affected by the dual particle/wave nature of both the electron and the photon. As a result of this, the precision of measuring electron's momentum and the precision of measuring its location are unavoidably traded off. Bohr extended this point in his overall account of the quantum states as defined by the supporting hypotheses. As Camilleri (2009, 94) warns, "We must not fall into the trap of reading Heisenberg's thought-experiment independently from the operational context in which it was proposed" to start with, and we should also understand it in the context of its amalgamation into Bohr's master hypothesis. Thus, "in referring to the velocity of a particle as we have here done repeatedly, the purpose has only been to obtain a connexion with the ordinary space-time description convenient in this case" (Bohr, 1928, 583). This description is heuristically suitable for the scattering experiments, but it cannot be generalized: "As it appears already from the considerations of de Broglie mentioned above, the concept of velocity must always in the quantum theory be handled with caution" (ibid.).

Thus, the uncertainty principle explains why the two supporting intermediate hypotheses, taken together, imply that the basic terms figuring in the everyday framework of the physical phenomena underlying experimentation (the first phase of the inductive process), along with the understanding of the physical world in classical physics, are not inherently connected to them. In fact, Heisenberg stated that eventually "the uncertainty relations were just a special case of the more general complementarity principle" (Heisenberg 1968, 106). In any case, the principle enabled assimilation of Heisenberg's and Schrödinger's accounts into supporting hypotheses of the emerging complementarity hypothesis, by outlining their key limitations in a precise fashion.

*

Bohr explicitly announced his complementarity account in his Como lecture of 1927,[4] and published that account in *Nature* in 1928 (Bohr 1928). This piece is a clear and succinct illustration of his methodological approach and all of its aspects: the exact role of individual experiments and data in hypotheses formation, supporting intermediary hypotheses stemming from them in a partial manner, and induction of the general hypothesis via an intermediary hypothesis. Bohr offered a detailed case for the unavoidability of both the particle and the wave approaches. He outlined a detailed experimental context and then suggested intermediary steps, echoing Heisenberg's and Schrödinger's proposals for general understanding. He discussed their limits, via the principle of uncertainty, and concluded by anticipating complementary features of the two. The piece is pretty much an exhaustive summary of all the experimentally driven points and criticisms we will revisit here.

Bohr's article published in 1925 (Bohr 1925), which we have discussed earlier in relation to the scattering experiments, was a precursor of the complementarity approach. In it, Bohr outlines all the steps he took before he explicitly formulated his master hypothesis. He gives a fairly detailed account of the key experimental findings, the supporting and constructive intermediary hypotheses drawn from them, and a hint of what a master hypothesis could look like. These experimental particulars and their selections were crucial both for Schrödinger's attempt to generalize his wave mechanical account and for the cautioning by Bohr that led to the master hypothesis. As we have seen, Schrödinger and Bohr disagreed on the low-level experimental hypotheses—that is, on how exactly to "cut up" relevant particulars. The resolution of the controversy over relevant particulars and their exact meaning—and thus the controversy over the justification for generating certain higher-level hypotheses, although not as protracted or as visible as the controversy over the justification for spectral analysis—was crucial. The scattering of light and matter performed by Compton and Simon and by Geiger and Bothe, using different experimental instruments and techniques, fairly quickly generated the grounding experimental hypotheses—or "accounts of experiments," in Bohr's parlance—that ultimately led to Bohr's account of complementarity as the master hypothesis. It did so via supporting intermediaries of matrix mechanics derived by Heisenberg, and wave mechanics developed by Schrödinger, as well as through the crucial constructive versatile intermediary hypothesis of the "uncertainty principle," devised by Heisenberg.

The main focus of Bohr's 1928 paper in *Nature*, however, is on establishing the master hypothesis of complementarity via the uncertainty principle, as a versatile constructive intermediary hypothesis that mediates two supporting hypotheses. Other previous steps outlined in his 1925 paper are retraced, but the presentation is organized with this central goal in mind. Bohr reflects on the methodological aspects in the context of this central goal, though not as much as he does elsewhere, in the passages discussed in previous chapters. Yet the bottom-up structure of the process is obvious here as well, and the details are worked out methodically, while dilemmas (paradoxes) and their resolutions, starting with those at the level of experimental particulars are at least indicated, if not discussed in some detail. The bottom-up (inductive) and hypothetical nature of the process is transparent, and the relationship between the experiments and the hypotheses is characterized in these terms. The physicist repeatedly "concludes," "draws," and "infers" from experiments in order to formulate intermediary hypotheses.

Despite the focus on the emerging master hypothesis, this presentation is not unrelated with Bohr's deeper view of how theory is generated (as discussed previously); quite the opposite. Bohr is interested in analyzing the actual experimental process and what is observed all along. Even his post hoc theoretical analysis does not veer away from the experimental situations into mathematical abstractions, or abstract models of physical phenomena. Rather, he repeatedly returns to them—for example, when the nature of observation is discussed in the context of performing an experiment, or when he develops a post hoc theoretical description involving quantum concepts. In contrast, Schrödinger's papers rely on toy models, refine wave-mechanical formalism, and rarely dwell on experiments alone. Meanwhile, Heisenberg's work uses the general discreteness of states in experiments as a starting point, but focuses on developing mathematical formalisms of matrix mechanics. Bohr's result was distinct, as was his method.

10: METAPHYSICAL PRINCIPLES AND HEURISTIC RULES

Building on the previous analysis, this chapter offers a unified under-standing of Bohr's approach which led to complementarity. It shows how his hands-on epistemological and methodological standpoint, which I propose to grasp with a few heuristic rules, was in stark contrast to Schrödinger's methodological stance. Analysis of the forms of expression in Bohr's milestone 1928 *Nature* paper provides the primary evidential and interpretive basis for this demonstration, and helps to address some of the misleading criticisms and misunderstandings of Bohr's account.

*

Throughout his dialogues with Heisenberg and Schrödinger, one of Bohr's main worries was that, given the way their hypotheses were pursued and argued—that is, as intended master hypotheses—the generalizations could have stifled development of the theory rather than serving as inter-mediary hypotheses true to the experimental context. This was especially true of Schrödinger's pursuit. In fact, the less productive way of pursuing scientific exploration by conveniently cherry-picking experimental particu-lars is likely to gratify the experimenter's favorite intuitions and metaphys-ical preferences while hindering the construction of a master hypothesis supported by a comprehensive record of the experimental context. As Ba-con had already pointed out, a person engaging in the pursuit of such an approach, in "anticipations," "merely brushes experience and particulars in passing" (Bacon 2000, 37, xxii). This approach thrives on the biases that guide the gathering of specific features of a particular experimental setup, without the biases being tempered by having the particulars put together with those selected in other experimental setups.[1]

In this less productive way, to borrow Bacon's point, the metaphysical preferences and theoretical biases of various sorts unjustifiably anticipate the general account by hastily narrowing down the experimental basis of the inductive process. The development of hypotheses anticipates a particular general account that is in unison with particular metaphysical preferences, while it only selectively takes into account various experimental particulars. Owing to metaphysically framed coherence, the common "intuitions" about the subject matter, hypotheses of that kind are initially almost guaranteed to elicit wide acceptance, the details and intricacies of the experimental context being sidelined in the process. Such "anticipations," as Bacon labeled them, "are gathered from just a few instances, especially those which are common and familiar, which merely brush past the intellect and fill the imagination" (Bacon 2000, 28, xxviii). The hasty selection results in a "settled truth."

Something of this sort is an unavoidable stage in complex scientific debates, and in the development of theories such as quantum mechanics. Yet these anticipatory claims were assimilated in a wider inductive process, resulting in a more adequate theory. The systematic and careful ascent to the master hypothesis set aside ready-made metaphysical presuppositions and ontological models of what reality should look like. It was guided by Bohr's central aim of a comprehensive grasp of the experimental context.[2] As a mediator in the division of labor in the quantum physics community, he was toning down those physicists who pursued their preferred accounts of microphysical states too hastily.

Schrödinger's early insistence on the wave-mechanical approach to quantum phenomena as being superior to the alternatives is a good example of such an approach. The contrasting roles played by Bohr and Schrödinger, and their different methodological approaches, clarify the context and the outcome of the controversy that emerged between the two. As we have seen, in contrast to many other physicists involved in the dialogue, Schrödinger explicitly insisted from the very beginning on the primacy of clear metaphysical ramifications as the bottom line of inquiry in physics. In light of that, it is quite understandable that Schrödinger was the most vocal critic of Bohr's approach, and that Bohr pushed back forcefully. Ultimately, Schrödinger had a restricted role to play in the broader effort pursued by the overall community.

Some authors (Beller 1999; Bitbol 1996) have argued that history would have taken a substantially different course if Schrödinger had managed to bypass the pressure of the so-called Copenhagen school, as the interpretation of quantum states organized around Bohr's views, among other things, was eventually labeled. This way, so the argument goes, he would have been allowed to freely pursue a philosophically more acceptable interpretation of quantum phenomena. Since quantum physics had just started, this could have been significant and would have affected the way the experiments and formalisms developed.[3]

It is important to understand that the debate between Bohr and

Schrödinger was not simply an "academic debate" on the preference of specific metaphysics associated with already established theory, though philosophers often like to portray it that way. Nor, as we will see shortly, was it a case of underdetermination of two distinct interpretations (motivated by opposed metaphysical and epistemological preferences) by an already widely accepted and formally unified theory. There was no underdetermination of any kind, as there were two ontologically (in terms of postulated entities and properties), empirically, and formally distinct approaches—namely, matrix mechanics and wave mechanics. The two approaches were equivalent only in a very restricted sense (Perovic 2008; Muller 1997a, 1997b, 1999).

In any event, as it should be clear by now, when Bohr was formulating his master hypothesis of complementarity, he aimed first and foremost to take into account a sizeable body of experimental work that supported the apparently substantially different hypotheses of Schrödinger and Heisenberg. By aiming at "doing justice to the different experimental facts," he aimed at nothing other than inducing an account of experimental particulars, an induction alleviated by moderate skepticism of the existing proposals to be generalized. In his 1928 paper he uses a characteristic formulation whenever he suggests that the relationship between the experimental basis and the intermediate hypotheses is established through such considerations. The experiments, he says, "find expression" in the second stage of drawing general hypotheses in either Schrödinger's wave-mechanical or the quantum-corpuscular account. He finally recognizes both as supporting intermediate hypotheses of the complementarity master hypothesis. Thus, "the experiences" of light interference "have found expression in the wave theory of matter" (Bohr 1928, 584), while "the conservation of the energy and momentum during interaction between radiation and matter, as evident in the photoelectric and Compton effect, finds its adequate expression just in the light quantum idea put forward by Einstein" (ibid.).

In such a process, the physicist should not prematurely explain away discrepancies in two substantially different sets of experiments by postulating entities evidently prompted by just one of them. The reason for pronouncing a hypothesis a general one that encompasses both sets of experiments should not be strictly, and especially not primarily, metaphysical. Thus, the physicist should not forcibly interpret the tracks as patterns of ionization in the scattering experiments performed in the cloud chamber—the interpretation actuated by wave-like appearances in the light interference experiments. The ontologically enticing wave-mechanical approach seemed to be applicable generally, and it certainly swayed Schrödinger. But in fact it was convincingly shown to be valid only for the angle of ninety degrees of the incoming light (Kidd, Ardini, and Anton 1985, 643), a fact that could not escape the ardent and perceptive arguers in the Institute for Theoretical Physics. It was not clear, then, that the scattering performed under nonstandard angles could be understood as using Mott's approach or an analogous one, nor especially whether it ought to be extended to the opposite cluster of experiments.

The logic of wave-mechanical ontology prompted such an interpretation and offered the formal tools to refine it, but the details of the experimental state of affairs gave Bohr reason to pause. In his judgment, such an approach would not acknowledge the details of the scattering experiments with enough rigor. It was quite possible that the tracks were patterns of ionization, which physicists saw as traces of rays solely because of their classical-mechanical treatment habits. Nor was this at odds with the appearance of a wave front hitting the screen in the interference experiments. However, the physicist might instead, and just as convincingly, interpret the traces using the hypothesis derived from another set of experiments—specifically those with spectroscopy, wherein the particlelike entities purportedly enter the spectrometer and leave behind well-segmented lines of spectra. Presented with such a choice, Bohr suspended judgment while listing the possibilities in a precise manner as "imperfect axioms," to use Bacon's term: intermediary, experimentally based hypotheses with a restricted grasp. The wave-mechanical and light-quantum hypotheses were helpful as long as they were provisional.

Bohr's emerging idea of a new master hypothesis was devised upward through the experimental context. It did not commit to the existing concepts as given. Instead, it constructed new concepts, the central one of which was "complementarity," by applying the old ones in a limited domain. The above-mentioned "finding of expression" of two sets of experimental hypotheses in two different intermediate hypotheses creates a dilemma and a paradox, both in this earlier instance and in the case of the experimental hypotheses that led to Bohr's model of the atom. Yet the paradox is deeper in the 1928 paper, when the concept challenged is "space-time coordination"—that is, the understanding of individual physical events as continuous spatiotemporal causal regularities and the common ground of classical physics. It was a challenge to Schrödinger's commitment to Boltzmann's way of approaching physical systems (including his understanding of the statistics of distinguishable states; i.e., wave modes in the quantum case) and the continuity principle he followed. Accordingly, the term *observation*, which is essentially tied to such an understanding, becomes problematic if used as a generalized abstract notion in post hoc theoretical considerations.[4] Although "energy and momentum are associated with the concept of particles, and hence may be characterized according to the classical point of view by definite space-time co-ordinates" (Bohr 1928, 581), "a further departure from visualisation in the usual sense" (ibid., 582) is inescapable in the sense of descriptions of experiments and regular language augmented with classical concepts. Thus, "an unambiguous definition of the state of the system is naturally no longer possible" (ibid.).

In a nutshell, in his 1928 paper Bohr summarizes the production of all the experimental results that led to his old atomic model, then the new ones that question it, and then the intermediate hypotheses of wave mechanics and the quantum-corpuscular account that introduce novel theoretical abstract terms with novel meanings. All these are back-and-forth attempts to formulate intermediate hypotheses based on the experimental hypotheses.

*

Meanwhile, Schrödinger was desperate to reinterpret the results of the scattering experiments in accord with his ontological motivation. He rejected Born's probabilistic interpretation of wave equation because he thought it adhered to quantum jumps and discontinuity. Relentlessly pursuing an "imperfect hypothesis" as a general hypothesis, as increasingly Schrödinger did in private (Joas and Lehner 2009, 349), may not be the best plan, as it becomes entangled in singular issues, thus requiring the postulation of singular, restricted ideas that cannot contribute to the understanding of the overall experimental context. Crucially, Mott's general approach is satisfying with respect to restricted measurements, at least as far as the experiments performed in the cloud chamber are concerned. Yet, for the Compton-Simon experiments an explanation along similar lines is satisfying only as long as the incident wave attacks the electron at an angle of ninety degrees (Kidd, Ardini. and Anton 1985). The explanation of scattering at other angles, predicated on the wave-mechanical hypothesis, requires amendment via impromptu interpretations and concepts hardly applicable to different experimental contexts. In fact, finally, Schrödinger acknowledged this in his paper on the Compton effect (1927b, 35), arguing that his wave-mechanical approach could not handle the phenomenon in individual processes satisfyingly. He gave up the pursuit of the approach after that.

A set of lower-level hypotheses prompts the development of singular restricted hypotheses connecting concrete experimental particulars. Such hypotheses ought to be brought into agreement later at a general level in the form of intermediary and master hypotheses. This is why it is not promising to opt for a candidate for the master hypothesis if the candidate is applicable to selective experimental particulars, even though it may have intuitive appeal. Such a hypothesis can become a general hypothesis only if it is somehow "forced" into the other relevant experimental contexts. Bohr's master hypothesis was premised on his usual attitude that it ought to arise from the overall experimental context.

In fact, in his 1928 paper Bohr provides an explicit and rather standard presentation of Schrödinger's hope that wave mechanics is a general account of quantum phenomena (Bohr 1928, 585–86). The following sentence in the section on wave mechanics and quantum postulate compares the two intermediate hypotheses: "In fact, wave mechanics just as the matrix theory on this view represents a symbolic transcription of the problem of motion of classical mechanics adapted to the requirements of quantum theory and only to be interpreted by an explicit use of the quantum postulate." The clear limit of the attempt to generalize wave mechanics is Schrödinger's "hope of constructing a pure wave theory without referring to the quantum postulate" (Bohr 1928, 589). More specifically, the "hope" that squarely downgrades it to being a supporting intermediate hypothesis is the quantum-corpuscular hypothesis drawn from the scattering experiments. Bohr argues that the "fulfilment of the claim of causality for the individual processes, characterized by the quantum of action, entails a renunciation as regards the space-time description" (ibid.). The individual states cannot be regarded as aspects of an inherently continuous space-time wave where such seemingly individual events are interference patterns. Rather, they are genuine discontinuities or quanta, becoming apparent in the light/matter and matter/matter interactions. Bohr says, "Only in this limit can energy and momentum be unambiguously defined on the basis of space-time pictures" (Bohr 1928, 584), that is, according to the demand for continuity. "For a general definition of these concepts, we are confined to the conservation laws, the rational reformulation of which has been fundamental problem for the [Schrödinger's] symbolical methods" (Bohr 1928, 584). What Bohr seems to be pointing out here is that, first, the proper general use of the concepts of momentum and energy in quantum physics is inevitably limited by the lower-level experimental hypothesis upholding the conservation laws of the values of the momentum and energy. Second, it is not clear how this hypothesis can be adequately reinterpreted within the bounds of Schrödinger's account, because any such attempt of interpretation (i.e., "rational formulation") would have to avoid treating momentum and energy as inherently noncontinuous conserved values. In a nutshell, instead of dwelling on such attempts of reinterpretation, Bohr suggests that the wave-mechanical hypothesis adequately accounts not for individual processes, but only for an ensemble as superposition of the states. The atomic states as characteristic vibrations of such a hypothesis are indeed useful, but only when the interactions of the particles in the atom are disregarded (Bohr 1928, 587). Moreover, the wave-mechanical hypothesis is removed from possible individual observations, and space-time coordination cannot be individualized, as Stern-Gerlach's experiment "brings out" (Bohr 1928, 588). Thus Bohr concludes that "a general conse-quence of the superposition principle," the foundation of Schrödinger's attempt to generalize the hypothesis, is "that it has no sense to co-ordinate [it] as phase value to the group as a whole, in the same manner as may be done for each elementary wave constituting the group" (Bohr 1928, 588).

Now this does not eliminate the reality of the wave packets postulated by wave mechanics. Bohr is more precise—and understanding this is cru-cial to understanding the nature of the complementarity thesis. It is not that the wave-mechanical account reduces to merely a formal instrument to account for relevant phenomena.[5] It has to be put in its proper place in the master hypothesis, which means it has to be taken seriously as an indication of the properties of microphysical systems. After all, "these "things" should not be thought of as objects describable with values of physical quantities like energy, momentum and position in the way we would expect on the basis of classical mechanics" (Dieks 2017, 305). Thus, Bohr defines them within the experimentally set limits: "In judging the possibilities of observation,"—that is, what could be observed as local and separable individual states, given the nature of observations—"it must, on the whole, be kept in mind that the wave mechanical solutions can be visualised only in so far as they can be described with the aid of the con-cept of free particles" (Bohr 1928, 587). Yet this is inadequate in quantum mechanics, since we either have quantum particles bound in interactions or a superposition of individual states. The term "free particles" cannot be used in either case.

Mara Beller and others argued that Schrödinger was accused of con-servatism because of his innovative standpoint. Actually, the reason for the accusation was his relentlessly traditionalist and metaphysically mo-tivated insistence on a particular "imperfect hypothesis." In a typically balanced manner, Bohr pointed out the limitations of the hypothesis to

Schrödinger without diminishing its substantial contribution as a limited intermediary hypothesis.

In fact, the method of balancing seemingly completely discrepant theories that accounted for certain separate domains of experimental results in a similar way had been pursued by Sommerfeld almost three decades earlier in his habilitation thesis. His account of diffraction aimed to synthesize Gustav Kirchhoff's results for radiation in a limited region with Poincaré's results for high-energy cases such as X-rays, derived by a completely different method. With methods much like Bohr's, though in a different experimental context, Sommerfeld was primarily seeking not a fully consistent model but a more tenuous account. The construction of the account moved from electromagnetism and the insights Kirchhoff treated as intermediary hypotheses, adequately induced from limited lower-level experimental hypotheses, to the treatment of these intermediary hypotheses by adequate tools of mathematical physics, and then back to suggestions for the tests of novel experimental hypotheses (Seth 2010, ch. 5). Sommerfeld stated, "As our theory casts each as suitable approximations of the exact formula, it forms a bridge between the two theories and apportions to both of them their restricted regions of validity" (Seth 2010, 146).

This part of Sommerfeld's work was, conceptually speaking, more akin to Bohr's complementarity phase than to the "symbiosis" of electromagnetism and Planck's quantum in Bohr's building up of the model of the atom. Sommerfeld in fact used the term "complementary" to characterize the relationship between the two intermediary hypotheses in the latter case (Seth 2010, 154), but Bohr might have picked up the term from him to address the former. In any case, the method was certainly not a complete novelty in the circle of quantum physicists.

Finally, even if eventually Schrödinger or any other physicist could have had adequately devised the wave-mechanical approach as a general account, Bohr erred on the cautious side, looking for an adequate long-term strategy. If Schrödinger managed to force his approach on the community, then Heisenberg, whose account was perceived as rival to Schrödinger's, would likely have been perceived as the loser, and his approach perhaps even abandoned.

*

A similarly illustrative and condensed example of Bohr's approach to a problem in physics that clearly exhibits all stages of the inductive process is also found in his later work on the fission of uranium. The work features his usual dissection of a problem into two stages and the development of lower, intermediary, and master hypotheses. This allows him to approach the problem differently than others and to resolve it, as in his previous major breakthroughs. In this case, and in the case of his milestone achievements "it was Bohr who saw with clear vision and intuition simple relations where others found themselves faced with the confusion of data and results" (Rozental 1968, 158).

Bohr initially sifted through numerous pictures, made in the cloud chamber at the Institute for Theoretical Physics, of uranium placed on metal foil. In comparison to previous pictures of alpha particles, these new ones exhibited some completely surprising curvatures of outgoing tracks and small branches emanating from them. Bohr induced a hypothesis based on these tracks. In this case, the alpha particles that leave completely different tracks in other processes, collide with very light electrons of the atoms of the gas, while the influence of the electrons on the tracks in those other experiments is not perceptible. Bohr went on to develop a more comprehensive account of the fission of uranium, and published it in *Physical Review* (Bohr and Wheeler 1939).

Once we have analyzed Bohr's approach and the role it played in devising his main contributions, as well as its immense influence on his contemporaries, we can try to sum up the process of deriving hypotheses in the form of a few simple heuristic rules. The derivation of lower hypotheses from the actual experiments will largely depend on the nature of actual experimental particulars; but once the experimental context is apparent, there seem to be some fairly transparent rules at work. These outline the boundaries of the inductive process, allowing us to formulate it. Although these heuristic rules can be traced across Bohr's work, including his published papers, notes, and correspondence, once again the document that unequivocally demonstrates them at work and showcases all elements of his method is his 1925 piece in *Nature* anticipating complementarity. The suggested rules below ought to capture the inductive process and give some rough but explicit guidelines to Bohr's thinking.[6]

First, clear and distinct experimental limits should be placed on existing and emerging theoretical frameworks, or more precisely, on the understanding of physical properties and processes. Thus, even though the quantum states can be understood as waves, this understanding is limited by insights from some experiments (e.g., scattering and collision experiments, experiments with spectral lines) as much as it is warranted by other experiments (e.g., light interference). This means we need to acknowledge the wave aspect of quantum states but not pronounce that quantum states are waves, or anything else that may follow from such a strong proposition.

Second, synthesizing theoretical accounts that seem opposed in light of particular metaphysical presuppositions can be beneficial in explaining known and experimentally examined phenomena and predicting new ones. This was something Bohr demonstrated in the use of the wave-mechanical approach and Heisenberg's approach, where he identified the advantages and shortcomings of each in accounting for particular phenomena. He also demonstrated it in certain features of his model of the atom. Bohr deemed the wave and quantum-corpuscular accounts complementary rather than mutually exclusive: they could be deemed exclusive only if the scientist stuck to the metaphysical presuppositions that seemed to underlie them. Thus, Bohr treated the wave-mechanical approach as complementary to Heisenberg's treatment, despite Schrödinger's insistence that the continuity of physical processes and objects, which Heisenberg and Bohr's discontinuous quantum jumps violate, was the baseline of any approach to quantum phenomena.

Our analysis of the complementarity approach from a hands-on epistemological and methodological standpoint should demonstrate that this was not primarily, if at all, an obscure metaphysical view. It has unjustifiably received harsh criticism (Maudlin 2018; Orzel 2005; Beller 1999; Bub 1974) because so many critics misunderstand its role and carelessly conflate it with later much more general views that Bohr formulated loosely, or that others attributed to him through the Copenhagen orthodoxy concoction, which we will turn to shortly. Those interested in metaphysical novelty would walk away disappointed when reading Bohr's views in the first three decades of the twentieth century, as his main goal was certainly not to build any such view in his account of microphysical phenomena. If physicists like Planck, Einstein, or Schrödinger pursued "the physics of principles" (Seth 2010, 3), and Sommerfeld pursued "the physics of problems" (ibid.), then Bohr pursued the physics of heuristic principles, with the heuristics driven by the rules outlined here.

*

Even though Schrödinger admitted he was defeated, he never abandoned his hope of reviving a version of his account as general. For example, in his 1927 paper on the Compton effect he acknowledged that his wave-mechanical interpretation of microphysical phenomena was inadequate, but he continued to emphasize his reading of Louis de Broglie's account as an argument for the plausibility of his wave-mechanical view. He argued that the rays of light could be described in terms of de Broglie's waves and the outcomes correspondingly calculated as differences in frequencies (Schrödinger 1956). The general idea was that the frequency could be ascribed to any "particle" to which definite values of energy were normally ascribed. Thus, in the scattering process, an incoming beam of light with a frequency $v1$ interacts with the free electron, exciting it to the frequency $v2$ of the recoil electron, which is equal to the difference between the former and the latter. To satisfy the requirement of the wave-mechanical description, given the formula $hv1 - hv1' = hv2 - hv2'$, the frequency of a vacuum of zero must be introduced. Schrödinger pursued this basic idea in explaining Franck collisions and the photo-effect (ibid.). Such a description turned out to be another way of making the case for the corpuscular nature of atomic interactions, albeit an unwitting one.[7] The coordinates appearing in the wave function are nothing but particle coordinates in three-dimensional space, a criticism put forward by Heisenberg (1955) and Born (1953) in the 1950s.

It seems, however, that Schrödinger was aware of this difficulty very early on (Scott 1967, 75–79; Beller 1997) and his search for a satisfying resolution anticipated the so-called second quantization, which allows the quantization of the electromagnetic field, where interactions of three-dimensional particlelike charges can be understood as the interactions taking place in the n-dimensional manifold. In other words, the method of second quantization enables us to understand the wave function, not necessarily as a function of three-dimensional coordinates, but rather as a function of the n-dimensional space of n coordinates—where expressions about the nature of the beam of light become expressions of the n-level excitations of the three-dimensional vacuum state, the frequency of vacuum being zero. Indeed, in a letter to Bohr following the October 1926 meeting, Schrödinger outlined desiderata for an account closely resembling such a view: "Perhaps the radiation dumping, the reaction on the system from the wave emitted by itself, should still be taken into account in a way quite different from what I originally thought, . . . through coupling to another system, the 'ether,' possessing a continuous spectrum of eigenvalues from zero to infinity" (Schrödinger in Bohr 1972–2008, vol. 6, 13).

In the same letter to Bohr, Schrödinger revealed his philosophical motivation for pursuing the view that motivated the entire program of 1926:

> I have not yet any definite ideas in this direction, and I should not impose my phantasies on you. What I vaguely see before my eyes is only the thesis: Even if a hundred attempts have failed, one ought not give

up hope of arriving at the goal, I don't say through classical pictures, but through logically consistent conceptions of the true nature of the space-time events. It is extremely likely that this is possible (ibid.).

The full-blown development of the idea of the second quantization started with the paper Dirac published in 1927 (Dirac 1927), but it is quite possible that Schrödinger missed Dirac's main point when he expressed his "phantasies" to Bohr in 1926. In his later accounts of the issue, he became increasingly aware of its importance for his own ideas.

Critical developments of the second quantization, especially those related to the conservation of energy and momentum laws, began with quantum electrodynamics (QED). Moreover, the continuity of the problems in Poincaré and Hendrik Lorentz's classical theory of electrons exhibited in Dirac's initial version of QED cast doubt on the ontology of quantum mechanics. These developments might have been a crucial encouragement for Hugh Everett, who developed his many-worlds interpretation essentially along the lines of Schrödinger's later ideas.[8] Once a viable physical expression of the n-dimensional manifold of electromagnetic interactions became available, interpreting the results of the "alternative" outcomes as real made sense. The n-dimensional manifold as the "space" of the events provided a physical background for the idea of the "reality" of the simultaneity of the outcomes and the explication of the measurements in terms of relative states. In a manner more reflective of Schrödinger's theoretical pursuit than Bohr's concern with descriptions almost directly read from experimental results, Everett attempted to solve the problem of measurement by proclaiming the reality of all the "alternative" outcomes provided by formalism.

11: NEW FORMALISMS
AND BOHR'S ATOM

The experimentally based inductive-hypothetical context extends into the ways novel formalisms of matrix mechanics and wave mechanics were understood to be equivalent following Schrödinger's milestone proof in 1926. That is the central topic of this chapter. The equivalence was not a purely mathematical exercise on two established interpretations of quantum states, as it is often and rather anachronistically portrayed. It was, in fact, understood in a domain-specific way, strongly shaped by the experimental context of each approach: the formalisms were neither conceived nor applied nor judged independently of the experimental context. The understanding of equivalence prompted a gradual transition from the model of the atom of the old quantum theory to a more adequate and comprehensive account of complementarity, and it also added another crucial limiting feature to supporting theories (hypotheses), along with the uncertainty principle.[1]

*

The community of quantum physicists fairly quickly reached a wide agreement that two competing formal accounts of quantum phenomena—Heisenberg, Born, and Jordan's matrix mechanics and Schrödinger's wave mechanics—were equivalent. Prior to this, the formalisms and their grasp of relevant phenomena had been perceived as substantially different, a distinction motivated in part by the mathematical techniques they employed. Matrix mechanics was an algebraic approach manipulating matrices, while wave mechanics depended on a partial differential wave equation. Moreover, the formalisms were initially applied to two distinct sets of experimental results. Matrix mechanics was successful in treating discrete appearances of spectral lines, and was then applied to the experiments with electron scattering. The initial applicability of wave mechanics to the experiments with light interference was soon extended to the energy values in experiments with hydrogen atoms. All this led to the general perception that the two accounts were distinct. When Schrödinger (1926e) published his paper arguing for the equivalence of the two, this was perceived as a major stepping-stone in the development of the theory.

Yet Schrödinger never provided a full-fledged proof of mathematical equivalence that a mathematically keen mind would welcome. Mathematical equivalence would mean that the mathematical structures of the two formalisms were isomorphic, but Schrödinger's paper contained only a preliminary attempt to prove this. He had a different goal; he aimed to prove a domain-specific equivalence with respect to the domain of Bohr's model of the atom—or, more precisely, to its particular features. At the time, Bohr's argument for a substantial convergence of the seemingly opposed and, in light of novel experimental results, revised theories was well on its way. Although mathematical equivalence of the formalisms did not seem out of reach, neither Schrödinger nor any other physicists who developed similar proofs intended to fully explore that possibility. The key reason was the priority they understandably gave to demonstrating the domain-specific equivalence of two supporting hypotheses (the domain being Bohr's model of the atom). Exploring mathematical relations between two formalisms was not as immediately important. As will be discussed later, Bohr perceived this "softer" equivalence as a decisive step in establishing a new master hypothesis. In fact, as Dieks (2017, 303) correctly noted, "Bohr's views are more intimately connected to the mathematical structure of quantum mechanics than usually acknowledged and reflect salient features of it."

*

There has been a tendency among historians of this episode in the development of quantum mechanics to simplify the situation and state that there was an agreement on the empirical evidence. What was left then, the argument goes, was to demonstrate equivalence of the formalisms. Yet, as noted above, the formalisms were neither conceived nor applied independently from the experimental context. The supposed empirical equivalence was not an *explanandum* and the supposed mathematical equivalence the *explanans*; far from it. The relationship between the experimental context and wave and matrix mechanics was complex. In fact, the physicists were trying to understand how exactly to apply the newly developed formalisms to particular experiments. Right after the meeting between Bohr and Schrödinger, according to Heisenberg,

> Bohr and his co-workers in Copenhagen became more and more concentrated on the central problem in quantum theory: how the mathematical

formalism was to be applied to each individual problem, and thus how the frequently discussed paradoxes, such as e.g. the apparent contradiction between the wave and particle model, could be cleared up. . . . We did not know how this mathematics should be used to describe even the most simple experimental situations such as e.g. the track of an electron in a cloud chamber (Heisenberg 1968, 104–5).

Moreover, the correspondence among the members of the community clearly shows that the physicists continued discussing the applications of the formalisms and their meaning well into the late 1920s, after Schrödinger's proof and the proofs of the others became available.

The formalisms, or "symbolic methods" as Bohr (1928, 581) labels them, enticed physicists to attempt to generalize wave mechanics and a quantum-corpuscular account supported by matrix mechanics.[2] These attempts exposed the limits of those hypotheses, the limits Bohr had formulated with the help of the uncertainty principle. "Harmonizing" the two within these limits was the goal of Bohr's master hypothesis. He says, "Only together [they] offer a natural generalization of the classical mode of description." That is, the classical mode of description could be generalized at a higher level of hypothesis only within the complementarity framework. With hindsight, we can state that "the principle of complementarity does not belong to the formal level of the theory but to the interpretive level" (Kauark-Lite 2017, 68) for the purposes of understanding its particular aspects, but the development of both was intertwined with and led by the same motivation.[3]

Schrödinger's proof, like Bohr's account of complementarity, was hammered out despite rather than thanks to an agreement on the experimental particulars. This is what made these achievements truly insightful in the eyes of the physics community, and what should make them admirable today. In general, prima facie, the two accounts were perceived as disparate since, as we have seen, on the whole they were straightforwardly adequate only for different aspects of the overall experimental context. Although each account individually was in clear agreement only with specific experimental particulars, there was a subtle formal agreement at the level of these two intermediary supporting hypotheses. In his 1929 Chicago lectures, Heisenberg explicitly "drew attention to the fact that the 'wave' or 'particle' features of the matter and radiation were brought out sharply by different experimental arrangements" (Camilleri 2009, 80). The proof aimed to demonstrate that they could be said to account for the overall experimental context.

In his proof paper, Schrödinger stated the following:

Considering the extraordinary differences between the starting-points and the concepts of Heisenberg's quantum mechanics and of the theory which has been designated "undulatory" or "physical" mechanics, and has lately been described here, it is very strange that these two new theories agree with one another with regard to the known facts, where they differ from the old quantum theory. I refer, in particular, to the peculiar "half-integralness" which arises in connection with the oscillator and the rotator (1926e, 45).

First, the "rotator case" in the above passage, which refers to the experimentally probed concept of quantization of orbital angular momentum, is perhaps the best indication of the lack of anything like straightforward empirical equivalence. Bohr's model of the atom accurately predicted the spectral lines of the Balmer series by introducing the quantized angular momentum of the electron. These values correspond with the rotational frequencies of the electron that Schrödinger introduced through an analogy with Bohr's model. As I have pointed out, Heisenberg took discrete values of the spectral lines as his starting point in developing the matrices accounting for these same values. Schrödinger (1926a, 30) admitted that his wave mechanics could not account for Balmer lines as straightforwardly as matrix mechanics could. The disagreement on this particular experimental result prompted him to seek common ground between the two approaches.

Schrödinger assumed that this failure of wave mechanics was only a matter of technical details (Schrödinger 1926e, 57), as previously noted. He set out to demonstrate a deeper equivalence and thereby establish the plausibility of his own approach by constructing the proof. He argued that this particular success and that of matrix mechanics reflected neither epistemological nor ontological advantages. Given the existing experimental particulars, he stated, we could not tell whether what the spectral lines indicate is applicable to individual corpuscular-like interactions of radiation with the matter (i.e., with the spectroscope), or whether the discrete values are the result of the way wave packets, rather than individual quantum corpuscles, interact with matter. The scattering experiments, following the first series of Compton's experiments and Ramsauer's (1921) earlier experiments, addressed precisely this issue.

Second, in the above quoted passage in the introduction to his proof, Schrödinger (1926e, 45) referred to the case of the oscillator, a special case of this wave-mechanical treatment of radiation. The way the atoms emitted energy, as very special oscillators, converged with the way that Heisenberg accounted for them. Schrödinger initially conceived the atom as a charge cloud, in contrast to Bohr's early model of an electron as a particle orbiting around the nucleus. Schrödinger did not adequately account for radiation of the atom, but Bohr did. Specific energy states corresponding to the allowed orbits were observed in the experiments with spectroscopy. Understood against the experimental background, the electric density of the cloud appeared different at different places but in a "constant fashion," in agreement with Bohr's model. In response, Schrödinger came up with the key idea of the cloud vibrating in two or more different modes with different frequencies. This, he said, would account for the radiation in corresponding energy states of the atom. In other words, the atomic states at different energies were characterized as eigenvibrations, and these were accounted for by eigenvalues of the wave equation. The radiation, then, is the atom emitting the wave packets of only certain energies, and this corresponds to frequency conditions in Bohr's model. Schrödinger also stuck to the view that the classical electromagnetic theory accounts for the manner in which atoms radiate; different values of radiating energy are achieved by the expansion in space of the emitted wave-packet of a certain energy. This assumption was accepted in Bohr, Kramers, and Slater's theory of radiation of the atom, as they attempted to deal with the continuous transfer of energy at the level of individual states. Schrödinger's (1926a, 1926b) initial major interest was the agreement between energy values arrived at by wave mechanics and those predicted by Bohr's theory (Jammer 1989, 275). This led him to consider the connection with matrix mechanics. The initial agreement between Bohr's atomic model and Schrödinger's wave mechanics on the nature of the atom's radiation is an essential element of the motivation for the proof.

Thus, it is not plausible that the physicists understood the oscillator—the coinciding energy values for the hydrogen atom—and the rotator "toy cases" as strong evidence of the empirical equivalence of matrix mechanics and wave mechanics. These cases were only indirectly relevant to the relations between matrix mechanics and wave mechanics, as the understanding was set against the background of Bohr's model of the atom. At the time Schrödinger wrote his proof paper, he as well as others knew that despite the initial agreement of his theory with Bohr's results with respect to the energy states and radiation, and the apparent advantage of Heisenberg's mechanics for spectral lines, the issue could potentially be decided only by exploring further experimental particulars, probing "individual radiation processes," and indirectly testing his own assumption about the vibrations of the atom.

Finally, the general background assumptions of Schrödinger's and Heisenberg's approaches and the limits of Bohr's model were probed experimentally by the second wave of scattering experiments tackling the nature of individual interactions. Schrödinger and Bohr initially perceived the results differently. Schrödinger perceived the results of the experiments as a caution to his attempt to generalize wave mechanical approach. But unlike Bohr, Heisenberg, and most of the community soon after, he was not entirely convinced until 1927 (Mehra and Rechenberg 1982, 138) that the results of these new experiments unequivocally demonstrated the discontinuous nature of interactions of matter and energy at the microphysical level. As he constructed his proof, the dilemma, or rather trilemma, had been addressed experimentally; but it remained unresolved as far as he was concerned.

Thus, as Schrödinger and others were devising their proofs, no one was certain whether or to what extent either of the two formalisms accounted for the observed properties of microphysical processes; nor did anyone know whether either was indispensable. Matrix mechanics and wave mechanics were certainly thought out and "designed to cover the same range of experiences" (Jammer 1989, 210) as master hypotheses accounting for the overall experimental context, but it was not firmly established in 1926 that either did so or in what way. What Schrödinger's proof and the proofs of other physicists addressed was not conceptual and formal discrepancy puzzlingly supervening on the full-blown empirical equivalence. Rather, Schrödinger aimed to show that, despite their selective accounting for the experimental context, there was a possibility that two approaches were equivalent within the specific domain of Bohr's atom, such equivalence being predicated on the agreement between eigenvalues and Bohr's energy levels. This agreement turned out to be demonstrable with rather simple manipulations of both formalisms. Schrödinger announced his intention right before developing his proof: "In what follows the very intimate inner connection between Heisenberg's Quantum Mechanics and my Wave Mechanics will be disclosed." In his second, related paper (sometimes labeled a "communication"), he more timidly stated that matrix mechanics and wave mechanics "will supplement each other" (Schrödinger 1926b, 30). He focused on their comparative advantages: "I am distinctly hopeful that these two advances will not fight against one another, but on the contrary, just because of the extraordinary difference between the starting-points and between methods, that they will supplement one another and that the one will make progress where the other fails."

Yet there was also a secondary goal, clearly on the back burner of the overall project, on which unfortunately the philosophical community and part of the physics community has since focused its attention, slowly and anachronistically perceiving it as the primary goal of the proof. Schrödinger stated at the beginning of the proof, right after he announced his main goal, that "from the formal mathematical standpoint, one might well speak of the identity of the two theories" (ibid.). This line, if taken out of the context of the structure of the proof, and if read without paying attention to the previously quoted formulations on the close intimacy of the approaches or Schrödinger's comments on their differences in the second "communication," could make us see the secondary goal of the proof as the primary goal—or even the only goal. Following the period of early developments of quantum mechanics, the proof and its intentions were customarily misinterpreted in this way.

To be fair, Schrödinger vacillated between the significance of these two goals before he put them on paper, as did other physicists in constructing their proofs. Another passage in the proof paper says that the goals and the nature of the proof are ambiguous (Schrödinger 1926e, 57–58). And in a letter to Wilhelm Wien, dated March 1926, he wrote that "both representations are—from the purely mathematical point of view—totally equivalent" (Mehra and Rechenberg 1982, 640).

*

Yet relying on selective explicit statements on the vacillating goals of the proof from correspondence and the paper itself comprises a limited sort of analysis. It is quite clear that the agreement of wave mechanics with Bohr's model of the atom was the key motivation for the development of Schrödinger's proof. It led to the crucial step in the proof: the construction of matrices based on the eigenfunctions. As Gibbins nicely puts it, "Schrödinger in 1926 proved the two theories equivalent . . . at least as far as the stationary, or stable-orbit, values for dynamical variables were concerned" (1987, 24). In fact this should not be surprising, as both matrix and wave mechanics were constructed against the background of Bohr's model and were attempts to improve and finally to replace it.

Bohr's old model steadily turned into a useful intermediary hypothesis used in the attempts to construct a new, more adequate master hypothesis. The proof in this sense was a transitional move from old quantum theory to a new master hypothesis. While Bohr's model had been changing since its inception, the importance of stationary energy states permitted by quantum rules in understanding quantum phenomena remained intact. And it became clear to what extent this core of the model remained insightful once matrix and wave mechanics were fully developed and the proofs of their equivalence were devised. This was a springboard to the construction of the proof and the transition to new quantum mechanics.

Practically speaking, Bohr's correspondence rules were the only available guidelines for the initial construction of matrix mechanics, initially conceived as little more than an improved version of Bohr's general account based on the model of the atom and the CP, and relying on the results drawn from the experimental context. Heisenberg (1925) did not initially use matrices but linear arrays of Fourier expansion. Once Heisenberg, Born, and Jordan produced the advanced version of matrix mechanics as a fully independent method of assessing the microphysical states (Born, Heisenberg, and Jordan 1926; Jammer 1989, 221), Bohr's model turned out to be only a first approximation of it.

In 1927, Lorentz pronounced at the Solvay Congress: "The fact that the coordinates, the potential energy, etc., are now represented by matrices shows that these magnitudes have lost their original meaning, and that a tremendous step has been taken towards increasing abstraction" (Lorentz in D'Abro 1951, 851). This was an exaggeration. Pauli applied matrix mechanics to the hydrogen atom and illustrated the method's independence from old quantum theory (Mehra and Rechenberg 1982, 656–57). Yet, like Schrödinger, he understood that the fundamental features of quantum phenomena characterized by matrix mechanics agreed with some key features of Bohr's model—most importantly, its core feature of stationary states, despite the novelty of matrix mechanics. This insight motivated Schrödinger to write the proof paper, since the agreement of wave mechanics and matrix mechanics in the proof was predicated on the key agreement between wave mechanics and Bohr's model.

*

If we replace the parameter of energy in Schrödinger's equation, the equation will have a solution. If a differential equation contains an undetermined parameter and has solutions only when particular values (so-called eigenvalues or proper values) are assigned to the parameter, the solutions of the equation are called eigenfunctions. Now, the solution in this particular case determines the amplitude of the de Broglie wave, and the eigenvalue, or the energy, determines the frequency of the wave. The chosen eigenvalue and corresponding eigenfunction determine the mode of eigen vibration. Schrödinger's solution of the hydrogen atom eigenvalue equation of his first and second communications resulted unexpectedly in Bohr's energy levels. Bohr characterized the situation in 1927 in the following way, pointing out the obvious agreement with his model: "The proper vibrations of the Schrödinger wave-equation have been found to furnish a representation of electricity, suited to represent the electrostatic properties of the atom in a stationary state" (Bohr 1972–2008, vol. 6, 96).

The entire argument for the advantage of the wave-mechanical approach over the matrix mechanics in the second communication was predicated on this agreement. And generally speaking, Schrödinger was greatly impressed by this newly discovered agreement, which raised the question of the system's apparently discontinuous nature, imposed on an essentially continuous approach of wave mechanics by quantum conditions. In 1926, while discussing the rotator case, Schrödinger noted the agreement between matrix mechanics and wave mechanics in quantum energy levels, clarifying the "half-integralness" as yet another point of agreement: "The intervals between the levels, which alone are important for the radiation, are the same in the former theory. It is remarkable that our quantum levels are exactly those of Heisenberg's theory" (1926b, 31). Others were equally impressed. For instance, Gregor Wentzel set out to examine this agreement with a new wave mechanics approximation method (Jammer 1989, 275–76).

At this point, wave mechanics was established as a methodologically independent treatment of microphysical states: "We have a continuous field-like process in configuration space, which is governed by a single partial differential equation, derived from a principle of action. This principle and this differential equation replace the equations of motion and the quantum conditions of the older 'classical quantum theory'" (Schrödinger 1926e, 45). Yet in light of this newly discovered agreement, it was not obvious that the independence of wave mechanics, like that of matrix mechanics, was not merely a methodological independence of a supporting hypothesis.

Given that wave mechanics and Bohr's model agreed with respect to the eigenvalues and stationary energy states, the main question the proof addressed was whether wave mechanics and matrix mechanics agreed with respect to eigenvalues and, thus, to stationary states as well. The proof was expected to prove the non–ad hoc nature of wave mechanics' assumptions and to have epistemological significance, something initially doubted by Heisenberg and others in the Göttingen school, including Schrödinger himself, because of its inapplicability to the spectral line intensities. The formal relations between mathematical structures of the formalisms were an explicit yet secondary concern. Although the main goal of the proof may seem modest to those educated in the tradition of philosophy and logic that equates understanding with reduction to some kind of mathematical-logical formalism, we ought to understand the goal within the context of the endeavor and its unknowns at the time. The importance of nailing down the "intimate connection" between matrix mechanics and wave mechanics was only vaguely apparent, and might have turned out to be insignificant if the independence of the two theories turned out to be more fundamental, especially if one were dispensable or had a very limited reach within the experimental context.

The "intimate connection" between wave mechanics and matrix mechanics is demonstrated by the construction of suitable matrices from eigenfunctions, the exact point of agreement between Bohr's model and wave mechanics. In fact, the structure of the proof becomes apparent when we realize that its purpose was to demonstrate the significance of their agreement with Bohr's model. The model accounted for a limited equivalence within the domain of this model—that is, a subatomic constitution. And it certainly did not directly or simplistically assume the experimental context and alleged empirical equivalence behind the agreement of the formalisms.

*

The proof paper consisted of the introduction, from which I have quoted Schrödinger's general characterizations, followed by three distinct parts developing the proof.[4]

Schrödinger points out a preliminary connection between matrix mechanics and wave mechanics early on in the proof, noting the key limiting feature of his attempt in the form of quantum conditions. He does this by explicating the background conditions of the correspondence between the use of matrix and wave mechanics, as limited by the quantum conditions of Bohr's model: "I will first show how to each function of the position and momentum-co-ordinates there may be related a matrix in such a manner that these matrices, in every case, satisfy the formal calculating rules of Born and Heisenberg (among which I also reckon the so-called 'quantum condition' or 'interchange rule')" (Schrödinger 1926e, 46). The interchange rules, the limiting parameter in Bohr's model—that is, its quantum rules—correspond to the analysis of the linear differential operators used in wave mechanics. Thus, in a very particular and limited sense, any equation of wave mechanics can be consistently translated into a corresponding matrix of matrix mechanics. Constructing suitable matrices from eigenfunctions established the "inner connection" between matrix mechanics and wave mechanics. This was the key to the proof's success, as both Schrödinger and the rest of the community saw it. It provided a unidirectional argument for the ontological equivalence in the context of Bohr's atom.

Next in his proof, Schrödinger replaces the suitable matrices with eigenfunctions of his wave equation: solving this equation is equivalent to diagonalizing the matrix H. Moreover, the Heisenberg-Born-Jordan laws of motion (Born, Heisenberg, and Jordan 1926) initially derived purely from matrix mechanics (Jammer 1989, 221), are satisfied by "assigning the auxiliary role to a definite orthogonal system, namely to the system of proper functions of that partial differential equation which forms the basis of my wave mechanics" (Schrödinger, 1926e, 46). At this point the main goal of the proof is achieved: the construction of matrices from eigenfunctions. Schrödinger was so confident in this that he later said he "might reasonably have used the singular" when speaking of matrix mechanics and wave mechanics. In a letter to Pauli on 8 June 1926, Heisenberg emphasized: "The great accomplishment of Schrödinger's theory is the calculation of matrix elements" (Pauli 1979, 328). Although Heisenberg's sense of the epistemological superiority of matrix mechanics over wave mechanics may have contributed to his attitude, the remark seems to show his understanding of the main goal of the proof.

Only now does Schrödinger turn to the secondary goal of the proof, merely stating in passing that the equivalence "also exists conversely." He never makes an effort to demonstrate this. Instead, he provides a vague idea of how we might proceed in proving this sort of mathematical equivalence. In the introductory section of the paper he explicitly emphasizes the importance of constructing matrices from eigenfunctions, while he vaguely hints at offering "a short preliminary sketch" (1926e, 47) of a derivation in the opposite direction, on a par with the attempt to explain the relativistic context of the wave equation in the last section of the paper. After achieving the main goal, he even labels the second part of the proof a "supplement to the proof of equivalence given above," and qualifies it as "interesting."[5] Proving the isomorphism (or S-equivalence; see Muller 1997a, 1997b) that required the proof of reciprocal equivalence clearly was not a top priority.

Now, achieving a proof of the mathematical equivalence of matrix and wave mechanics would have made sense only if a full-blown empirical equivalence was established. Otherwise, given that the ontological and methodological status of wave mechanics and matrix mechanics was tentative, a much more pressing and tangible issue of the relations between matrix mechanics, wave mechanics, and Bohr's model was aimed at by means of a "softer" construction of matrices from eigenfunctions. The more tangible equivalence in the key feature of Bohr's model did not require bidirectional derivation to prove isomorphism.

Moreover, probably in response to his debate with Bohr, Schrödinger deflated his high expectations, and this crucially contributed to the focus on the "soft" derivation and the goal in the proof. In fact, the assertive tone and insistence on the exclusiveness and superiority of wave mechanics over both old quantum theory and Heisenberg's approach, very explicit in his first communication (Schrödinger 1926a) and somewhat toned down in the second (Schrödinger 1926b), disappear in the proof paper. The tone is defensive. Schrödinger cautiously argues that wave mechanics may have the same epistemological significance as matrix mechanics, and treats the part of the paper dealing with this issue as secondary. Even if Schrödinger was at first undecided as to the main goal of the proof, following his debate with Bohr, he and the quantum physics community embraced it as defined by its limited domain-specific goal.

Two years after the publication of this seminal work, Bohr continued to discuss the application and meaning of wave mechanics in his correspondence. In a letter to Schrödinger dated 1928, the key issue was still the nature of the agreement of wave mechanics and matrix mechanics with the features of Bohr's model (now firmly demoted to a supporting hypothesis) and an assumption that stationary states were a limiting condition on the applicability of wave mechanics:

> In the interpretation of experiments by means of the concept of stationary states, we are indeed always dealing with such properties of an atomic system as dependent on phase relations over a large number of consecutive periods. The definition and applicability of the eigensolutions of the wave equation are of course based on this very circumstance. (Bohr 1972–2008, vol. 6, 49)

Finally, similar proofs constructed by others around the same time established similar aims. Thus, in a letter to Jordan, Pauli talked about "a rather deep connection between the Göttingen mechanics and the Einstein–de Broglie radiation field" (Mehra and Rechenberg 1982, 656). He thought he had found "a quite simple and general way [to] construct matrices satisfying the equations of the Göttingen mechanics." The use of the key feature of Bohr's model was instrumental in this case as well. In fact, Pauli's attitude was strikingly similar to the complementarity view devised by Bohr in response to the same developments. After presenting the relations between matrix mechanics and wave mechanics in his letter to Jordan, he concluded that "from the point of view of Quantum Mechanics the contradistinction between 'point' and 'set of waves' fades away in favor of something more general" (Mehra and Rechenberg 1982, 657). Although the full-fledged empirical equivalence was not in place, the proof demonstrated that two seemingly disparate approaches could be features of a more general account of microphysical states and processes. Bohr certainly took that crucial message away from this episode.

Interpretations of formalisms, their application, and the experimental context were in flux—different aspects of the same effort. The experimental context, theory, and interpretations were interdependent, and they continuously informed and influenced each other. Drawing sharp lines between them just to generate neat distinctions and arguments that favor a particular view of the development of physics is historically unsubstantiated, and prevents us from fully understanding the nature of Bohr's endeavor.

12: COMPLEMENTARITY
ESTABLISHED AND APPLIED

In this final chapter on Bohr's role in the emergence of quantum mechanics, I flesh out the meaning of complementarity in terms of its treatment of the limits of different intermediary hypotheses, and I explain the subsequent reactions of the quantum physics community. As I argue, complementarity was not confined to reconciling these hypotheses; it also helped provide some guidance in different approaches to quantum tunneling soon after Bohr developed his new master hypothesis. Although it did not have the breakthrough flair of Bohr's model of the atom, it has had a considerable and long-term influence on the experimentalist community.

*

The new master hypothesis Bohr devised tells us, then, that two supporting intermediary hypotheses most clearly expressed by Schrödinger's and Heisenberg's work (1) are incapable of further generalization in lieu of experimental particulars, as is precisely demonstrated via the uncertainty principle; (2) complement each other formally and in the overall experimental context; and (3) do so with respect to specific sets of experimental particulars.[1]

In principle, we might expect a satisfying hypothesis drawn from experiments to be analogous to the character of actual measurements and experimental particulars to the same extent as classical theories. However, this was not the case in quantum mechanics, nor was it the case in old quantum theory—and Bohr seems to refer to this when he states, "It is just this entirely new situation as regards the description of physical phenomena that the notion of complementarity aims at characterizing" (1935, 700). Nor do wave or matrix mechanical accounts suffice as a general account. An effective account has to encompass the experimental context, sorting out and synthesizing intermediary hypotheses. For example, an intermediary hypothesis stemming from interactions of light with matter states that microphysical properties exhibit particlelike nature. In several other circumstances its different aspects exhibit a holistic nature, defying localization and separateness, something that impressed Schrödinger and motivated his account. Hence, complementarity, synthesizing the two, is a general hypothesis based on those intermediary hypotheses, which are in effect its limit cases. Francis Bacon offered a convincing general description of the second stage of the experimental inductive process, the stage in which the master hypothesis is built. As noted earlier, we just need to replace his notion of axiom with our notion of hypothesis: "For the lowest axioms are not far from bare experience . . . the intermediary axioms which are the true, sound, living axioms. . . . And also the axioms above them, the most general axioms themselves, are not abstract but are given boundaries by these intermediary axioms" (Bacon 2000, 83, civ).

The classical notions used in "interpreting experience"—the two sets of experiments leading to the intermediate hypotheses, the "most elementary concepts" in the wave-mechanical and quantum-corpuscular approaches—have to be reinterpreted as complementary, as they cannot be taken to be inherently related. As Bohr tells us, "The complementary nature of the descriptions appearing in this uncertainty is unavoidable already in an analysis of the most elementary concepts employed in interpreting experience" (Bohr 1928, 581). The first stage—characterized by spatiotemporally coordinated, causally connected individual events, and individual, local, separable states—is heuristically generalized in the second-stage inferences, within a limit. Everyday experience, concepts of observations directly based on perceptions, as in experimental work, and their *simpliciter* extension into the theoretical framework are only partially possible in quantum mechanics ("a fragmentary application of the classical theories"; Bohr 1928, 584). These properties of regular observations are a heuristic conceptual tool in explaining quantum-corpuscular phenomena of scattering—the limit of which is demonstrated by the superposition principle.[2] This is why we cannot assign reality to agents of observations and observed systems separately (Bohr 1928, 589). Here, classical notions of isolated particles and radiation in free space are abstractions, but they indispensably augment accounts of experiments in lower-level hypotheses.

*

Having this summary of the overall argument in mind, we are now in a good position to handle the concerns about Bohr's allegedly mysterious statement that quantum mechanics is complete, implying that such joint completion of phenomena renders any single phenomenon incomplete in some sense. Carsten Held paraphrases and comments on this view: "In a clear quantum mechanical sense one well-defined phenomenon does 'exhaust the possibilities of observation,' while in another mysterious sense it is 'complementarity in the sense that only the totality of the phenomena exhausts the possible information about the objects'" (Held 1994, 886). In the beginning of the second inductive stage, the phenomena as experimental particulars are incomplete, as they have been gathered in the perceptually and conceptually biased isolation we discussed earlier. This stage aims at gradually producing a hypothesis that is all-embracing of gathered experimental particulars. What is more, it need not present an adequate image consistent with any ready-made metaphysical criterion and principles. It is adequate in that it introduces complementary physical features through intermediary "imperfect hypotheses"—wave-mechanical and quantum-corpuscular in this case.

Bohr states, "Just this situation brings out most strikingly the complementary character of the description of atomic phenomena which appears as an inevitable consequence of the contrast between the quantum postulate and the distinction between objects and agency of measurement, inherent in our very idea of observation" (Bohr 1928, 584). Thus, general metaphysical notions of individual states and the separability those states presume stem from regular observations based on sense perceptions. We inevitably use these notions at the first stage of descriptions. At the theoretical level, however, these states are used as only partial heuristics that back up an intermediate hypothesis. And at that level they are used along with, and are complementary to, the concept of inherently superposed states. The inherently superposed states as such are ultimately unobservable, in contrast to the states (individual and separable) of stage one of the experimental process. Both general accounts of the two kinds of states, individual and superposed, are thus partially utilized at the theoretical stage.

The complementarity approach was a provisional synthesis of different approaches to microphysical phenomena. To be more precise, it was a synthesis of intermediary hypotheses, each closely tied to the various segments of the rich experimental context. This approach was perhaps rather surprisingly minimal, but it was dependable. Various experiments tackled numerous related phenomena by using numerous experimental techniques, thereby eliciting distinct lower hypotheses that initially appeared inconsistent and, as such, appeared to be obstacles to a more general account.

Heisenberg gradually and reluctantly accepted Bohr's point as a reasonable master hypothesis after his initial misgivings about the necessity of using both accounts as inescapable supporting hypotheses. He also had misgivings about whether individual experimental results had to be accounted for by both supporting hypotheses taken together (Camilleri

2009, 77–82). His reluctance was motivated by his commitment to the quantum discontinuities that had led to the development of his theory, and by his strong opposition to the supporting wave-mechanics hypothesis. In his 1960s recollections he stated, "For me the essential point was that I had understood that by playing between the two pictures, nothing could go wrong. So I didn't object to playing with both pictures. At the same time I felt that it was not necessary. I would say it was possible but not necessary" (Heisenberg in Kuhn 1963, 21). Finally, and mostly because he concurred with the discrepant experimental situations emphasized by Bohr, Heisenberg accepted Bohr's account as a tentative but reasonable master hypothesis. The crucial bridging work, incorporating both supporting hypotheses into the master hypothesis, was the uncertainty principle: in particular, the gamma-ray microscope thought-experimental account of microphysical states in light of the Compton scattering. After a prolonged dialogue with Bohr, Heisenberg concluded that "in the observation of the electron position, the direction of the Compton recoil is only known with an uncertainty" (Heisenberg 1983, 84; see also Heisenberg to Pauli, Bohr 1972–2008, vol. 6, 19). This meant that the quantum-corpuscular supporting hypothesis had to be amended to incorporate wave mechanics.[3]

Jordan (1944, 131) accepted Bohr's complementarity hypothesis wholesale as a breakthrough in physics and philosophy, while Pauli (1980, 7) opined that quantum theory might as well be labeled the theory of complementarity. Eventually, the bulk of the community accepted the complementarity account, as it was amalgamated into the Copenhagen orthodoxy I will turn to in the next chapter. This fact on its own does not indicate the adequacy of the account, yet the rationale for and key objective of inductive methodology in general led to a tentative master hypothesis through a balanced and comprehensive assessment of the experimental results. It also steered experimental exploration and theoretical work in various directions.

The contention surrounding complementarity has persisted, however, and philosophers and historians of science continue to explore it. It is desirable to devise an account that agrees with traditionally accepted metaphysical principles and widely held "intuitions." Yet this objective is secondary to doing justice to the overall experimental context rather than to cherry-picked phenomena, and it is frequently not achievable. To those physicists and philosophers whose understanding is guided by received metaphysical notions, or to those who treat interpretation of experimental results as simply a selection between a couple of ready-made metaphysical principles, the result of a thoroughgoing induction from a complex experimental context will inevitably seem obscure or even incoherent, and they will try to refute it as such. Bohr's arguments seem puzzling, obscure, or inconsistent (Held 1994) if viewed solely in light of metaphysical preferences. They lose their aura of obscurity if we perceive them as results of the inductive methodology and its aims. Thus, we would certainly not expect that many professional philosophers or physicists nurturing particular metaphysical preferences would find the results of Bohr's inductive approach as enticing as, say, Schrödinger's account. What ought to be more agreeable to such judges, however, is the method producing the results. And the soundness of this method led even Bohr's fiercest critics to grudgingly accept those results. This was also eventually true of Schrödinger himself, as shown above.

The physicists who expected Bohr's master hypothesis to cohere with the standards they deemed intuitive found it disappointing. At one point, Einstein even labeled Bohr's approach "religion." And as we have seen, Schrödinger, until his encounter with Bohr, criticized his general approach as surpassing what he thought to be the limit of intuitive explanations of physical phenomena. Even though Einstein's and Schrödinger's ontological and epistemological commitments differed, they both argued that a general characterization of microphysical processes should postulate either waves or particles as basic, but never both.

The master hypothesis Bohr created may not be appealing if we assess it by the standards of ready-made metaphysical accounts. Yet, given the experimental context of the time, it is hard to see how anyone could have created a general account that would have been appealing in this sense. Any other solution on offer would have excluded some of the experimental particulars, and the pragmatic feature of Bohr's approach was that the scientist should err on the side of caution. The formalism was flexible and could be developed in many directions, but this flexibility was a virtue only when the experimental context was adequately accounted for. Otherwise, it could reduce to fancy formal and theoretical tricks, which in turn could be obstacles to further development. Schrödinger did not try to sell such tricks, yet his approach got stuck at the level of experimental details, not at the level of details of his formalism.

*

Dugald R. Murdoch (1987) interprets Bohr's complementarity as a form of pragmatism with a realist slant, making use of classical concepts within the limits of quantum principles. We can talk about waves, but only as they are limited by the quantum of action. In addition, quantum theoretical methods can be used only symbolically, and classical concepts cannot be interpreted in a purely realistic sense. Murdoch illuminates this aspect of complementarity by analyzing Kant's influence on Bohr through Høffding. Jan Faye (1991) also relies on the Kantian aspect of Bohr's philosophical approach but arrives at the opposite conclusion; he determines that Bohr was a staunch proponent of antirealism. And more recently, Bitbol (2017, 55–59) has used the machinery of Kant's philosophy to interpret complementarity.

These and similar, more or less vague philosophical ideas, or ideas wanting further clarification, may have been used at various points in Bohr's pursuit of physics (Heilbron 2013, 37). But there are many—perhaps too many candidates—to "explain" Bohr's choices. Bohr's assessment of the experimental situation was the basis of his choices and the foundation of his intellectual tools and ideas. Too often these attempts to pinpoint a particular influence of a particular philosophical idea are only vaguely plausible or informative, and read more as wishes that particular ideas were employed as intellectual tools. But we have no satisfactory way to find out if any of those truly were used in that way. Also, such analyses often rely on the intellectual musings Bohr practiced throughout his life, which are too far from any experimental context (e.g., Heilbron 2013, 37). It is not clear why we would try to connect them directly to his pursuit of a particular achievement in physics in the first place.

Either way, the inductive process Bohr pursed is inherently intersubjective. Thus, a division of labor in the scientific community is both inevitable and required for success. Various and often opposed stances are pursued and relentlessly advocated by physicists who formulate and follow hypotheses with a limited experimental grasp. Bohr understood his role in presenting the community with relevant experimental facts and imperfect hypotheses elicited from them in a partial way, while restraining from hasty preferences for any one of them. In other words, he avoided being compelled by the logic of any "imperfect" hypothesis that could serve as an argument for its generalization.

Bohr's inquisitive tone and his open-ended statements, often misinterpreted as confused, resulted from his moderate skepticism about ready-made ontology and its principles. In Schrödinger's case it was the continuity principle and the *Anschaulichkeit* requirement, while in Heisenberg's case it was the insistence on exclusively discrete states combined with the effectively antirealist, mathematically driven analysis of microphysical phenomena. In his conversations with Schrödinger and Heisenberg, Bohr listed various aspects as seemingly opposing accounts, as they were connected to the experimental results. Meanwhile he tried to find out how they amounted to a more general hypothesis. The benefit of such analysis is that the concepts used in imperfect hypotheses are refined as they are brought into a broader context of multiple experimental situations. This has a potential to eventually remove the stark contrast between imperfect hypotheses. The experimental particulars in such a case are connected by a tentative account (an intermediary hypothesis), and the physicist should refrain from accepting ready-made accounts that are experimentally unfit to be promoted into general accounts. The second wave of the scattering experiments was not an instance of crucial experimentation in physics in the sense that it proved the wave-mechanical hypothesis to be incorrect and the corpuscular hypothesis correct. It was crucial in the sense that it paved the way for a master hypothesis.

As we have seen with respect to Bohr's model of the atom, the final aim of such an inductive process is to devise a comprehensive hypothesis composed of the "rational utilization of all possibilities of unambiguous interpretation of measurements." Each possibility, such as the wave account or the quantum-corpuscular account, results from the interpretation of specific unambiguous particulars accounted for in the lower-level hypotheses. Each one, however, leaves out certain other relevant particulars, so the inductive process should use both to construct a more general hypothesis. Thus, two imperfect intermediary hypotheses were united into a general hypothesis of complementarity, which was an adequate and sufficiently comprehensive experimentally driven master hypothesis. In fact, as we have pointed out, exactly such construction of complementarity was already presented by Bohr in his 1928 paper published in *Nature*. In it he exhaustively listed all the relevant experimental particulars we have discussed, and demonstrated the advantages and disadvantages of wave-mechanical and discrete-states-based approaches within the overall experimental context. This paper went through perhaps the most revisions of all Bohr's papers, even though all papers went through an unusually high number of revisions. The revisions involved practically all collaborators at the Institute for Theoretical Physics (Klein 1968, 90), as the paper aimed to provide the entire inductive scaffolding of relevant experiments and the hypotheses elicited and supported by them. Indeed, if we want to grasp Bohr's vision of physics in practice and its result in his complementarity phase by reading a single work, his 1928 paper in *Nature* is the piece to read.

*

In a discussion of Bohr's complementarity account, Bitbol (2017, 47) recently summarized the common view that "physicists needed an expedient theoretical scheme and some guiding representations, whereas Bohr rather developed a general reflection about the epistemological status of such schemes, and stressed the limitations of representations in science." The latter part of the statement is certainly correct: Bohr indeed constructed his view as a gradual account of the limitations and advantages of accounts that ensued from the experiments. But as we will see, Bohr's complementarity met the expectation spelled out in the first part of the statement, though not as comprehensively as his model of the atom had previously done. The complementarity approach was more than a working hypothesis that satisfyingly encompassed the existing experimental context. It was also a useful and dependable guide for theoretical and experimental studies of unexplored phenomena, setting aside the fact that it assumed a life of its own in the amalgamation of the so-called Copenhagen interpretation.

A particularly telling example of this usefulness was early research on quantum tunneling. Bohr encouraged it around the time he was developing his complementarity approach, and this left a mark on different lines of early research. In general, as Jeremiah James and Christian Joas point out, the significance of the early applications of quantum mechanics—applications understood as use of a theory to explain a new phenomenon—has been overlooked by historians in their bid to understand the ramifications of the theory's acceptance (James and Joas 2015, 642n). In accord with their methodological warning,[4] the early work on the phenomenon of quantum tunneling may illuminate how complementarity was regarded and used in the community, and how its explanatory power and generality were tested.[5]

There is a small probability that a particle with less energy than is required to overcome a potential hill will "tunnel" through it instead. The likelihood of such an effect is very small, but it increases with the number of collisions. Such a phenomenon is naturally not allowed by classical mechanical laws, as the particle bounces off the barrier (figure 9).

In the initial stages of the discovery of this effect, Friedrich Hund (1927a, 1927b, 1927c) treated the pairs of atoms as double-potential wells, the dynamics of the atoms that compose molecules, and analyzed the so-

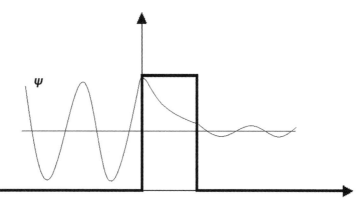

Figure 9. The particle wave can occasionally "tunnel" through a barrier requiring energy higher than the energy of the particle wave. Certain features of the quantum tunneling effect (e.g., the "tunneling time," the time it takes the particle wave to tunnel through the barrier) are still not fully understood.

called luminous electrons. In contrast, Lothar W. Nordheim (Fowler and Nordheim 1928, 1927) saw the thermionic emission of electrons and their reflection off metals as free particles' continuous behavior, arguing that the emission of electrons from a metal surface in a strong electric field was due to the tunneling effect. The breakthrough research of George Gamow (1928) and of Ronald W. Gurney and Edward U. Condon (1928, 1929) investigated the connection between quantum tunneling and alpha decay. Since Rutherford's experiments with the alpha particles, it had been obvious that the atomic nucleus consisted of a vast number of particles bound together by strong nuclear forces. But the experiments also demonstrated that the nucleus can absorb a low-energy particle of the same charge as those composing the nucleus, while radioactive elements can emit high-energy particles of much lower energy than the energy required for the particle to break out from the forces keeping the nucleus together. Gamow's model was based on the analogy between wave mechanics and wave optics. He assumed that the particles were very energetic, and that the barrier in the nucleus had to be between certain width sizes. Crucially, on the one hand, he concluded that the "leakage" of the matter-wave prevented the formation of sharp stationary states, but that on the other, the leakage was very small. He postulated the state at stake as quasistationary, where the matter-wave conserved probability by appropriate damping of vibrations captured by a complex energy expression Gamow had devised.

Early researchers also formulated a crucial feature of the effect: the time or velocity of tunneling through the barrier. They developed an equation for the mean time spent within the barrier by the packet of particles penetrating it. This aspect is still hotly debated, and there are currently four different approaches. The first approach relies on experimental measurements of a delay in the arrival of the reflected and transmitted wave packet. The second traces ticks of an "internal clock" via degrees of freedom of the barrier-particle system. The third traces an incoming particle's semiclassical trajectories using Feynman diagrams, Bohm's mechanics, or the Wigner distribution. The fourth analyzes the probability density within the barrier and the density of the incoming flux ratio (the dwell time).

But how is this historical path of the discovery and especially the current stage of research relevant to Bohr's pursuit of complementarity?

Bohr's 1928 *Nature* article mentioned the ongoing work of Hund on molecular spectra, work he in fact encouraged (Hund 1927a). But Bohr did not dwell on it, as the goals of the paper were not theoretical-speculative, and at that time tunneling was waiting for more substantial experimental efforts to shed light on it. This is why Bohr emphasized the classical aspects of the electron in the tunneling dynamics, and briefly outlined its limitations in the paper (Bohr 1928, 590). In fact, it seems he understood the effect as yet another important limitation, along with the scattering and collision experiments, of the attempt to generalize the particle-like aspects of quantum systems interactions.

Bohr's major contribution was not his direct penetration of the problem, but the influence his complementarity framework exerted on the research of the phenomenon, and still exerts today. In accord with that novel framework, Hund used Bohr's old atomic framework with orbital transitions as the framework of two bound states, but crucially amended it by the wave-mechanical account of the behavior of the potential well as a nonstationary oscillation of the state of superposition between these two states. The wave-mechanical framework of quantum systems was not fully embraced as a general account in this case, but was embraced in a limited sense the way Bohr's complementarity argument suggested it should be.

Gamow approached this quantum effect in the same manner as Bohr approached supporting hypotheses when contriving complementarity. His account synthesized isolated features as they were grasped by Hund and Nordheim into a comprehensive hypothesis. And this similarity in the methodological approach may not be a coincidence; Gamow was impressed by Bohr's work at the time (Gamow 1966, section 4) and was actively encouraged by him. (Casimir 1968, 110). Similarly, Gurney and Condon synthesized in a more formal manner Nordheim's continuous treatment of a barrier and Hund's bound-states treatment. Their account of molecular bonds was generally well received, but full-blown acceptance only came with work by Walter H. Heitler and Fritz London (1927). However, these two authors, in turn, relied on the previous work, especially Hund's.

It is hard to avoid the impression that features of the complementarity framework frame the current debate on the tunneling time. It is also hard to see how that could be avoided at this point. Thus, the results of the experiments with atomic and nuclear collisions suggest coincidence between the quasiclassical limit of quantum definitions of tunneling times and the usual classical-mechanical expressions (Olkhovsky et al. 2004). And although two approaches to the tunneling effect are essentially distinct in their grasp of it ("for discrete energy spectra the time analysis of the processes is rather different from the time analysis of processes corresponding to continuous energy spectra"; ibid. 168), the analysis of the time evolution of collisions based on semiclassical trajectories using Feynman diagrams yields the same results as Schrödinger's purely wave-mechanical analysis—an equivalence of sorts of the two distinct approaches. Generally speaking, the situation seems closely analogous to the context bridged by Bohr's complementarity, in which the physicists are dealing with theoretically distinct yet equivalent supporting hypotheses, each grasping part of the experimental evidence: the hypotheses that ought to be grasped with a comprehensive model. It may not be surprising, then, that in the spirit of Bohr's complementarity approach, experimentalists are explicitly keen on gathering and comprehensively assessing the experimental results, rather than prematurely trying to develop any one existing approach into a general account of tunneling time while sidelining others (Chiao 1998; Winful 2006). This is especially understandable given that the current subfield of quantum mechanics which focuses on the tunneling effect is a particularly good example of how the careful gathering and analysis of experimental setups works its way into experimental and theoretical hypotheses, and how various theoretically driven biases can affect the choice of experimental particulars. Bohr's complementarity principle certainly cannot resolve the puzzle of the tunneling time, but its methodological backdrop seems to be a backdrop of the debate.

PART 4

Aftermath

13: BOHR AND THE "COPENHAGEN ORTHODOXY"

But in the usual interpretation of the quantum theory, an atom has no properties at all when it is not observed. Indeed, one may say that its only mode of being is to be observed; for the notion of the atom existing with uniquely definable properties of its own even when it is not interacting with a piece of observing apparatus, is meaningless within the framework of this point of view.

—David Bohm (1957, 92)

It is a remarkable leap from the experimentally driven hands-on debates among the physicists in the 1920s, discussed so far, to the characterization of the "usual interpretation of the quantum theory" that David Bohm summarized in the above quotation some thirty years later. A few paragraphs earlier, in fact, he attempted somewhat cautiously to ascribe this same interpretation to Bohr, stating that a "similar point of view is indeed already implicit in Bohr's conclusion" (Bohm 1957, 92).

We have seen, however, that Bohr did not advocate an antirealist treatment of individual microstates, but rather saw it as a confounding of Heisenberg's partial theory. Another reason his attitude should not be surprising is that, at least until the 1930s, when interpreting microphysical states he was much too concerned with the intricacies of the experimental context, and with the induction of hypotheses from them, to even consider any general statements of the sort Bohm characterizes. How did such a strong antirealist account become "the usual interpretation of the quantum theory," and the interpretation that prominent physicists like Bohm thought was implicit in Bohr's own account?

First, Bohr's complementarity approach took on a life of its own soon after the debates of the second half of the 1920s were provisionally settled. Bohr's very general philosophical musings and speculations on the applicability of the concept in other areas of science that were removed from the experimental context and inductive method based on it coincided with this. In fact, after the basics of the theory were established in the mid-1920s, Bohr and others at the Institute for Theoretical Physics began to ask "whether one should seek a physical interpretation of the formalism along quite other lines than those hitherto considered by the Copenhagen group. . . . From now on the interpretation of quantum mechanics was the most important subject of our discussion" (Heisenberg 1968, 101). The discussion accorded with Bohr's vision of the master hypothesis as provisional and open for revision, yet always within the bounds of the existing experimental context. Due to Bohr's prominence in the community and the authority he commanded, his approach could not be sidelined no matter what anyone else said about microphysical states and quantum mechanics. Complementarity thus gradually turned into a kind of figure analogous to those in Rorschach psychological tests. In such a situation, physicists, historians, and philosophers too often project their own preferences regarding whether they want the major authority on their side, or aim at discrediting it. Bohr had become a victim of his own entirely justified success in the early development of the theory.

Second, and as might be expected, the community tried to lump various key aspects of the results of the debates in the 1920s and 1930s into an effective and simple conceptual narrative about quantum phenomena, including the formalisms in use, the key experimental results, and the complex understanding behind them. The "Copenhagen interpretation" or "Copenhagen orthodoxy" was thus gradually concocted; as Bitbol (2017, 47) correctly notes, it was "a mixture of elements borrowed from Heisenberg, Dirac, and von Neumann, with a few words quoted from Bohr and due reverence for his pioneering work, but with no unconditional allegiance to his ideas." Camilleri (2009, x) concurs, saying it "comprises a number of different viewpoints and philosophical positions." Understandably, the concoction was even less appealing to a metaphysically sensitive ear than the early products of Bohr's method before they became conflated with the rest of the emerging orthodoxy. "It is not necessarily linked with a specific philosophical or ideological position. It can be, and has been, professed by adherents to most diverging philosophical views, ranging from strict subjectivism and pure idealism through neo-Kantianism, critical realism, to positivism and dialectial materialism" (Jammer 1974, 87). This was especially beneficial to physicists, like Bohm, who needed a unified springboard to develop their own accounts, substantially different from the basic formal and conceptual elements of the quantum theory.

We should not find this development surprising or intellectually worrisome, however, even if we strongly disagree with what the development offers to the understanding of quantum mechanics. The scientific process is typically a series of provisional accounts; sometimes, but very rarely, it turns into a beauty contest among the various abstract theoretical accounts of phenomena of interest. And we also know why this is the case. The experimental context is typically wide and complex; grasping it as a whole is extremely challenging and takes long periods of gestation.

There have been a few attempts (Howard 2004; Beller 1999) to debunk the rather reticent "Copenhagen consensus" that emerged after the debates of the 1920s. Rather than a coherent interpretation created through the process of mutual agreement, it is deemed a result of coercion similar to the coercion that allegedly led physicists to converge on Bohr's approach to microphysical states. This is perhaps too harsh a judgment. The Copenhagen interpretation was only a stage, albeit a simplifying or even simplistic stage, in a gradual, wider, and thoroughly justified agreement. This brings us back to the attempts to devise proofs of substantial equivalence between matrix and wave mechanics: based on everything we know about the equivalence proofs, they must have served as solid bedrock for the emergence of this wide agreement, crucially contributing to both the complementarity account and the Copenhagen interpretation. However, in accord with the notion of the myth about the wide agreement, Frederik A. Muller (1997a, 1997b, 1999) deems the equivalence a myth as well. It is worth looking briefly into the details of his argument to reverse the misunderstanding of what sort of agreement physicists had accomplished at the time under Bohr's umbrella, and how the agreement was absorbed into the Copenhagen interpretation.

Muller argues that only the work done by the physicists in the 1930s, especially that of von Neumann (1932), provided a sound proof of the mathematical equivalence of matrix and wave mechanics. Schrödinger's famous proof, as well as the proofs of Dirac (1930), Carl Eckart (1926), and Pauli (1926) fell short of this objective. The agreement they prompted, the argument goes, was predicated on the misconception that the empirical equivalence was established, and that the mathematical equivalence had been successfully proved.

This view has further implications for the understanding of that period of development of quantum mechanics. It concurs with the idea that a wider agreement in the physics community at the time was thoroughly unjustified, as was the subsequent development in interpreting the theory predicated on it, which led to the complementarity approach and the Copenhagen interpretation. Without a successful proof, the agreement on Bohr's synthesis of wave mechanics and matrix mechanics, which Bohr and others thought successfully countered the exclusive commitments to either continuity or discontinuity, was seen as having been forced upon the community by the Göttingen group (Beller 1999). An alternative view is that a myth of the Copenhagen interpretation was built by deliberate or semideliberate misinterpretations of the history (Howard 2004).

In fact, the supposed myth of the equivalence based on Schrödinger's allegedly unsuccessful proof must have played a key role in these scenarios. If the agreement on the mathematical equivalence was unjustified, and the equivalence of the mathematical structures of the theories was the center of attention, then the theories should have been treated as distinct, and the complementarity approach should have been seen as an essentially baseless hodgepodge synthesis. The distinctness of the competing theories in the face of an unproved equivalence could not be a valid reason for the agreement against the arguments for either the wave-mechanical or the matrix-mechanical approach as suggestions for a general account of microphysical states and processes. Indeed, this train of thought seems to be inevitable: if the mathematical equivalence was not proved in the 1920s, then the theory did not favor the Copenhagen interpretation over Schrödinger's and Heisenberg's interpretations. An early wide agreement as constituting the beginning of the Copenhagen interpretation is puzzling at best.

As argued previously, however, the goal of the proofs was not what Muller thinks it was, and the development of formalisms, theory, and experimental context was inherently interrelated in a complex manner. The problem is that the analysis of the above-outlined sort is not firmly situated in the historical context, and as such it is bound to miss some of the key aspects of the debates among the physicists. Such historically impoverished analysis is limited, because it is predicated on certain models of scientific knowledge that focus the analysis almost exclusively on particular mathematical-logical aspects of the theory structure—in this case, matrix and wave mechanics as formalisms.[1] Once we focus our analysis on secondary, formal goals of the proofs of the 1920s, the agreement on the equivalence, the ensuing wider agreement on complementarity, and the establishment of the Copenhagen interpretation start to look like a string of myths or unsubstantiated viewpoints. Yet this is predicated only on leaving out the main goal of the proofs: to improve the existing understanding of the atom, in concert with the wider goal of developing an experimentally driven understanding of microphysical states and processes. The equivalence proved in the 1920s did not aim as high as the proofs in the 1930s. The goal of Schrödinger's proof may seem modest and its result mathematically less tractable in comparison, but it justifiably led to the gradual convergence of the community of physicists on Bohr's approach, and thence to a wider agreement. The roots of what became the Copenhagen interpretation lie in this domain-specific equivalence of matrix mechanics and wave mechanics; it certainly did not start as the manufacturing of consent among physicists and philosophers.

While the proofs of the 1920s were being developed and discussed, the physicists were still set on obtaining results and discussing the implications of novel experimental tests of the corpuscular and wave mechanical hypotheses. Once the results were deemed to provide a satisfying overall general account in the form of the complementarity approach, the development of quantum mechanics could enter the second phase, and the concerns about mathematical details, including mathematical equivalence of formalisms, could take center stage. Yet even von Neumann's proof of the 1930s could not settle the dilemma. Although "von Neumann's theory was a splendid achievement . . . it was also a precisely defined mathematical model, based on certain arbitrary, but very clearly stated assumptions concerning quantum theory and its physical interpretation" (Hanson 1963, 124). Crucially, the scattering phenomena exhibited in the second series of experiments with Compton's effect could not be adequately formulated from the point of view he developed at the time (ibid.). More importantly, when this second stage of the development of quantum mechanics began, partly led by the spirit of hammering out an acceptable "usual interpretation" of quantum mechanics, the commentators among the physicists and philosophers gradually equated Schrödinger's proof with the spirit of von Neumann's proof. Yet we should not confuse this equivocation with the actual goal of the proofs in the 1920s: namely, to lay the groundwork for a wide agreement solidified in the complementarity approach and then extended into the semipopular Copenhagen interpretation.

It is hard to fathom the notion of the "usual interpretation of quantum mechanics" as presented by Bohm in the above passage, when we have immersed ourselves in the debates of the 1920s. It is equally hard to see the connection between Schrödinger's ideas of the 1920s on the wave-mechanical nature of microphysical states and the role the wave function played in the context of these ideas, on the one hand, and the idea of the *collapse of the wave function* as a defining feature of the Copenhagen interpretation, on the other. It is especially puzzling to understand the usual idea of the collapse as the real feature of quantum interactions from the various points of view of the 1920s discussed to this point. The superposition of the states, accounted for by the wave equation, collapses upon measurement; that is, the state changes into one of the components of the superposition when the measurement is performed. This collapse is attributed to a special physical status of the act of measurement. This is an idea perpetrated after von Neumann's (1932) suggestion.

That this idea is hard to understand from the context of the 1920s debates nicely illustrates how remote the subsequent developments were from the early ones. It is another important reminder that we should be careful not to interpret debates of the 1920s, including their epistemological and methodological features, in an anachronous manner or by conflation. For instance, are measurements that result in collapse of the wave function physically real and recordable events? What is their ontological status in comparison with those states for which the wave function accounts? An argument for a collapse of the wave function as real may conflict with Schrödinger's account of the energy exchange as essentially continuous. It also opens up new dilemmas that did not emerge in the 1920s: Should we treat all state evolutions as governed by the linear Schrödinger equation, or, as Johansson suggests, is the "Schrödinger evolution . . . only appropriate when the system is isolated" (Johansson 2007, 94)?

Another broad view of quantum states that developed well after the debates of the 1920s, and which does not really spring from them in any obvious way, is the *decoherence view* (Paz and Zurek 1993). It states that the physical environment "measures" a quantum system by interacting with it, and stores the information of the measurement. Any physical system interacting with the measured system can play the role of the observer. This view is one of many views grounded in a now widespread understanding of the theoretical structure of quantum mechanics, reduced to little more than bare formalism in a preferred form, as distinct from the "measurement problem"—a term gradually introduced and widely used only after World War II. The theoretical structure is taken to account for the physical system prior to the moment of the measurement that interferes with it, as it were. What happens after that moment is open for interpretation, as is the issue of what sort of system exists prior to it. Either way, it is clear that the formulations of the theory after the 1920s substantially reformulated and refocused the debate on the microphysical states, and introduced new concepts and conceptual frameworks.[2]

14: BOHR'S RESPONSE TO THE EINSTEIN-PODOLSKY-ROSEN ARGUMENT

I know that no living person has looked so deeply into the actual abysses of quantum theory as the two of you, and that nobody else sees how necessary are completely radical new conceptions.

—Paul Ehrenfest, commenting on Albert Einstein and Niels Bohr[1]

A perceived paradox at the heart of the newly established quantum mechanics was advocated in the famous paper by Einstein, Boris Podolsky, and Nathan Rosen in 1935. The usual brief textbook summary of the argument (e.g., Howard 2007; Norton 2018) states that since particles must have definite observable properties, such as spin, even before any measurement is performed on them, quantum mechanics cannot be complete, as it does not account for hidden properties. This leads to the correlation of the spins of two particles—for example, two atoms leaving the molecules because of electrostatic force (the force that does not disturb their spins). Quantum mechanics gives us possible values of the property (of spin) that we choose when performing the measurement, but it cannot give us a full description of the system. The two particles are really separated and, in principle, the measurements can be performed on each particle independently. What quantum mechanics offers us is just a statistical account of particle ensembles, not of individual states and their properties which lie hidden in the fabric of physical reality and wait to be discovered by a new and better complete theory. As Bohr correctly summarized it: "Einstein here argues that the quantum-mechanical description is to be considered merely as a means of accounting for the average behaviour of a large number of atomic systems and his attitude to the belief that it should offer an exhaustive description of the individual phenomena" (Bohr 1949, 235).

As is well known, Bell's argument (1964), and subsequent experimental tests of it (Aspect, Grangier, and Roger 1982), demonstrated that if quantum mechanics correctly predicted the outcomes of measurements, then there was no set of hidden properties of the type that Einstein assumed to exist, that would be consistent with the set of such outcomes.

A striking aspect of the Einstein-Podolsky-Rosen (EPR) paper is its

forceful opening. It clearly states some very broad philosophical propositions that the authors claim ought to define any complete theory:

> In a complete theory there is an element corresponding to each element
> of reality. A sufficient condition for the reality of a physical quantity is
> the possibility of predicting it with certainty, without disturbing the sys-
> tem. . . . Any serious consideration of a physical theory must take into
> account the distinction between the objective reality, which is indepen-
> dent of any theory, and the physical concepts with which theory operates.
> These two concepts are intended to correspond with the objective real-
> ity, and by means of these concepts we picture this reality to ourselves
> (Einstein, Podolosky, and Rosen 1935, 777).

As Howard (2007) notes, "Einstein later noted that the separation principle is a conjunction of two logically independent assumptions, today
termed separability and locality, and he presented deep philosophical
premises for each. But the basic logic of Einstein's intended incompleteness argument remained the same." These stated claims to which Howard
refers, which are quoted above, are convincing if we subscribe to a very
particular philosophical realist standpoint. If we believe physical theories
are merely instruments for the reliable prediction of phenomena, we will
not subscribe to these postulates, nor will we demand that our theory com-
plies with them. Yet they set the tone and the motivation for the overall
EPR argument. It is certainly clear that Einstein's measure of a superior
theory was to provide agreement with physical reality via "secure" explanations, the terms used by Planck and Sommerfeld (Seth 2010, 187). Einstein
perceived Bohr's theory as merely "satisfactory" in its agreement with reality, and thus incomplete and open to revision.

This argument and the debate between Einstein and Bohr on the nature of quantum mechanics as a scientific theory that preceded the actual
paper have been the topics of numerous analyses, debates, exaggerations,
and controversies. Before I go on to discuss some important aspects of the
controversy, the following assessment of the debate seems fair: "Bohr still
emerges with the better arguments. . . . But Einstein's legacy is rehabili-
tated, his dissent being seen for what it was: principled, well-motivated,
based upon deep physical insight, and informed by a sophisticated philosophy of science" (Howard 2007).

The most discussed account of Bohr is the piece he wrote in response
to the EPR paper (Bohr 1949), even though there is plenty of other relevant material in his correspondence with other physicists. When Bohr

challenged the classical framework of physics while he was building his model of the atom, the actual experiments led him to do so. When Einstein challenged the general assumptions of quantum mechanics in the EPR paper and in the overall debate with Bohr, he did not have any experiments that led him to that view. He was a philosopher legitimately defending a particular viewpoint and anticipating theory and experiments that he expected would prove him right, or a physicist wedded to the view of physics as "theories of principle" as opposed to, in his view, the inferior "constructive theories" pursued by Bohr and others—a distinction Einstein drew in a *Times* article published in 1919. Bohr and Einstein did not belong to the group of physicists mathematizing physical problems to the extreme— "virtuosi" as Einstein labeled them. Both were seeking the principles behind physical phenomena, but their methods of seeking them followed different goals, and hence led to different kinds of principles. Both Bohr's challenge of classical mechanics and the EPR challenge of quantum mechanics are legitimate in science, but it is important to realize that they have very different natures. Bohr was deeply aware of this, and that was the main reason why he thought Einstein's challenge was wanting from the very beginning, despite its forcefulness: "In my opinion, there could be no other way to deem a logically consistent mathematical formalism as inadequate than by demonstrating the departure of its consequences from experience or by proving that its predictions did not exhaust the possibilities of observation, and Einstein's argumentation could be directed to neither of these ends" (Bohr 1949, 229).

Despite this fundamental weakness in Einstein's argument, Bohr's overall response was somewhat dull in comparison to the arguments he had devised in his previous work. This was unavoidable. He simply could not demonstrate in this debate what he was best at: deriving hypotheses from actual experiments. He could only summarize the recent history of the development of quantum mechanics, and repeat how experimental conditions give rise to complementarity as a response. The response is, however, perhaps the most succinct summary of these developments, and of the gradual development of Einstein's attitude towards quantum mechanics.

Bohr's opening summary is more convincing than the second part, in which he contributes some general philosophical musings he had started developing in the late 1920s, near the end of the period of building quantum theory and then quantum mechanics. As it is abstract and more in line with Bohr's mature general thoughts on complementarity, the second part is detached from the inductive process of "reading" the hypotheses from new experimental results. Here, Bohr's argument becomes as abstract as Einstein's. This whole aspect of the debate is detached from the actual experimental context, and is predicated on very abstract notions. It raised concerns about possible hidden variables, and anticipated the lines of attack on the problem, as it were, that became pertinent and were appreciated only in the late 1920s after the foundations of quantum mechanics were developed. Yet focusing on this part of Bohr's response in order to disparage complementarity and the role it played in the 1920s is a waste of time. The debate raised a very important question, but the EPR arguments for the incompleteness of quantum mechanics, as well as Bohr's response, remained fairly marginal to the main story of the development of quantum mechanics, almost like a long and prominent footnote to it, at least before World War II. The question only became central to the foundational discussions of the theory later, especially with the work of John Bell.

Although the first part of Bohr's response summarizing the development of quantum mechanics to date is predictable, it is effective because Bohr is explicit about how the EPR argument brushed over experimental particulars and the ways they were synthesized, so that it could make an abstract point. As noted earlier, Bohr never dismissed the need to understand individual physical states, the way Heisenberg did. But the experimental context led him to the view that such individual states could not be conceived in a classical manner. Thus, the theory inevitably led to the "inability of the classical frame of concepts to comprise the peculiar feature of indivisibility, or 'individuality,' characterizing the elementary processes" (Bohr 1949, 203). To this he added:

> Einstein was perhaps more reluctant to renounce such ideals than someone for whom renunciation in this respect appeared to be the only way open to proceed with the immediate task of co-ordinating the multifarious evidence regarding atomic phenomena, which accumulated from day to day in the exploration of this new field of knowledge. . . . In the following years general methods were gradually established for an essentially statistical description of atomic processes combining the features of individuality and the requirements of the superposition principle, equally characteristic of quantum theory (ibid., 206).

Yet the conviction about the nature of individual physical states was gradually amplified in light of incoming experimental results: "The paradoxical aspects of quantum theory were in no way ameliorated, but even emphasised, by the apparent contradiction between the exigencies of the general superposition principle of the wave description and the feature of individuality of the elementary atomic processes" (Bohr 1949, 207). Ridding the desired allegedly "complete" account of either the wave or the particle features, as Einstein and Schrödinger respectively anticipated, could not have been achieved.

As the final master hypothesis, "complementarity was suited to embrace the characteristic features of individuality of quantum phenomena" (Bohr 1949, 209). What preceded it was the stage of gathering experimental particulars in the only way humans can possibly do it: "It is decisive to recognise that, *however far the phenomena transcend the scope of classical physical explanation, the account of all evidence must be expressed in classical terms*" (ibid.; emphasis in the original). The master hypothesis, the considerations of the apparatus, and any given physical phenomenon in quantum terms came only after the classical observations were gathered. Appropriate hypotheses, no matter how unintuitive, were formed based on them.

Bohr also notes that the application of emerging formalisms to the experimental context was challenging: "The problem again emphasizes the necessity of considering the whole experimental arrangement, the specification of which is imperative for any well-defined application of the quantum-mechanical formalism" (ibid., 230). Once we are equipped with theoretical knowledge of this sort, we can choose how to set up the apparatus and how to perform an experiment that will capture the wavelike or the particlelike aspects of microphysical phenomena:

> In the quantum-mechanical description our freedom of constructing and handling the experimental arrangement finds its proper expression in the possibility of choosing the classically defined parameters entering in any proper application of the formalism. Indeed, in all such respects quantum mechanics exhibits a correspondence with the state of affairs familiar from classical physics (ibid.).

It was inevitable, albeit disappointing to people like Einstein, that "evidence obtained under different experimental conditions cannot be comprehended within a single picture, but must be regarded as complementary in the sense that only the totality of the phenomena exhausts the possible information about the objects" (ibid., 210). As became clear in the mid-1920s, light interference experiments, spectral analysis, and the scattering experiments were two sets of experiments leading to such conclusions. "Under these circumstances an essential element of ambiguity is involved in ascribing conventional physical attributes to atomic objects, as is at once evident in the dilemma regarding the corpuscular and wave properties of electrons and photons, where we have to do with contrasting pictures, each referring to an essential aspect of empirical evidence" (ibid.). The master hypothesis induced from the experimental work is, unfortunately for those committed to a particular kind of realism about physical states, such that "while the combination of these concepts into a single picture of a causal chain of events is the essence of classical mechanics, room for regularities beyond the grasp of such a description is just afforded by the circumstance that the study of the complementary phenomena demands mutually exclusive experimental arrangements" (ibid., 211). In effect, Bohr says that, given the experimental context, the best induction or rational generalization of the master hypothesis we can have is provisional, but that it can be substantially challenged only on the experimental basis and not on the basis of philosophical preferences. Accordingly, Einstein's challenge "in no way points to any limitation of the scope of the quantum-mechanical description" (ibid.), except as a weak challenge based on a rather very abstract argument without a novel experimental situation to substantiate it.

After this elaborate summary of the development of quantum mechanics and its induction from the experimental results, Bohr himself turns to a more abstract argument. The sober experimentally-minded statement of the limitation of the EPR challenge is extended with a more ambitious claim removed from the actual experimental context: "The trend of the whole argumentation presented in the Como lecture was to show that the viewpoint of complementarity may be regarded as a rational generalisation of the very ideal of causality" (ibid.) He elaborates on this strand of the argument by saying, "In this respect, quantum theory presents us with a novel situation in physical science, but attention was called to the very close analogy with the situation as regards analysis and synthesis of experience, which we meet in many other fields of human knowledge and interest" (ibid., 224). This is a reference to his general expectation that complementarity is a powerful concept with wide applicability and that, as such, it offers an epistemological alternative to Einstein's rather naive form of realism.

Bohr points, much as Einstein does, to the abstract philosophical viewpoints at the basis of his arguments. In perhaps the most forceful statement of this sort in the piece, he comments, "We are not dealing with an arbitrary renunciation of a more detailed analysis of atomic phenomena, but with a recognition that such an analysis is *in principle* excluded" (ibid.

229). This statement is ambiguous and can be read as a reinforcement of the master hypothesis, given the experimental context, or as a challenge of Einstein's realism from the viewpoint of Bohr's general epistemological standpoint. Bohr also turns to another long-standing intellectual preoccupation—in this case, the limitations of language in accounting for physical phenomena: "A precise formulation of such analogies involves, of course, intricacies of terminology, and the writer's position is perhaps best indicated in a passage in the article, hinting at the mutually exclusive relationship which will always exist between the practical use of any word and attempts at its strict definition" (ibid., 224).

The passages attest to what we have already characterized as Bohr's mature thought, detached from the actual experimental context and motivated by different, rather abstract concerns. He goes on at length to clarify them in this piece:

> A principal purpose of such parallels was to call attention to the necessity in many domains of general human interest to face problems of a similar kind as those which had arisen in quantum theory and thereby to give a more familiar background for the apparently extravagant way of expression which physicists have developed to cope with their acute difficulties. Besides the complementary features conspicuous in psychology and already touched upon, examples of such relationships can also be traced in biology, especially as regards the comparison between mechanistic and vitalistic viewpoints. Just with respect to the observational problem, this last question had previously been the subject of an address to the International Congress on Light Therapy held in Copenhagen in 1932, where it was incidentally pointed out that even the psycho-physical parallelism as envisaged by Leibniz and Spinoza has obtained a wider scope through the development of atomic physics, which forces us to an attitude towards the problem of explanation recalling ancient wisdom, that when searching for harmony in life one must never forget that in the drama of existence we are ourselves both actors and spectators (ibid., 236).

Understandably, many physicists did not have time for such musings and labeled them too philosophical. Similarly, Bohr the experimentally minded inductivist did not have time for Einstein's argument in the first part of the paper. Yet it was not unusual for the physicists at the time to put forward a particular philosophical assumption matching reasoning about a particular physical phenomenon. They were typically philosophically educated and versed, and were part of philosophical circles in one way or another (e.g. Einstein was an active member of the Vienna circle). This is why Einstein could make a forceful argument in the EPR paper in the first place. Bohr does the same here in the second part, although he made sure it belonged to the very end of the inductive process. His invocation of the double-aspect philosophical accounts is closely reminiscent of Mach's understanding of his neutral monist understanding of mind and body. The view postulated the substance as neither strictly physical nor strictly mental, but rather a synthesis of properties human observers encounter in perception and in their understanding of the physical world. Physical entities are not fixed, but are fleeting and dynamic in the way items in the perceptual experience are fleeting and dynamic. Bohr's account is akin to both Spinoza's and Leibniz's "psycho-parallelism," the aspect he emphasizes. Mach was not only an inspiration for the Vienna circle and for the logical and empirical positivism that developed from it, but also a tremendously influential philosophical figure among the physicists of Bohr's generation. He was in fact a physicist himself who was controversially offered a philosophy chair at the University of Vienna and accepted it. As Karl Popper put it, "Few great men have had an intellectual impact upon the 20th century comparable to that of Ernst Mach. . . . He influenced Albert Einstein, Niels Bohr, Werner Heisenberg, William James, Bertrand Russell—to mention just a few names" (Popper 2002, 151–52). In light of Mach's neutral monist account, Bohr's argument could hardly be interpreted as a return to obscure philosophy at the time; more likely, it was read as a predictable piece of philosophical thought that aimed at clarifying the account that the inductive process produced, especially the nature of physical entities, as only partially localizable and corpuscular, yet real.

The extension of the everyday language of observations used in the experimentation can go only so far in building higher-level hypotheses— only as far, in fact, as the intermediate supporting quantum-corpuscular hypothesis allows. Within such limits, Bohr said much earlier, "We find ourselves here on the very path taken by Einstein of adapting our modes of perception borrowed from the sensations to the gradually deepening knowledge of the laws of Nature" (Bohr 1928, 590). The final result, the complementarity principle, puts this epistemic aspiration into a more general perspective. Finally, as far as the general philosophical nature of the final results goes, Bohr states, "I hope, however, that the idea of complementarity is suited to characterize the situation"—a very cautious and tentative formulation of the goals of the master hypothesis—"which bears a deep-going analogy to the general difficulty in the formation of human ideas, inherent in the distinction between subject and object" (Bohr 1928,

590). The metaphysical duality statement comes at the very end again in the 1928 paper—in the top layer of drawing hypotheses.

An aspect of the 1949 piece that was perhaps more pertinent at the time, and certainly more useful in understanding Bohr's overall thought, is his clear dismissal of aspects of the amalgam that the Copenhagen interpretation had become. The EPR took it for granted: "This, according to quantum mechanics, can be done only with further help of measurements, by a process known as the reduction of the wave packet" (Einstein, Podolsky, and Rosen 1935, 779). This was hardly a view that Bohr advocated, as I explained in chapter 13 on the Copenhagen orthodoxy. He does his best to distance his views from another aspect of the Copenhagen Interpretation as well: the notion of the "disturbance of measurement." Indeed, it is hard to see where it would fit in the work that resulted in his breakthroughs, or even in his more general epistemological account, and why it would be necessary. Bohr says:

In this connection I warned especially against phrases, often found in the physical literature, such as "disturbing of phenomena by observation" or "creating physical attributes to atomic objects by measurements." Such phrases, which may serve to remind of the apparent paradoxes in quantum theory, are at the same time apt to cause confusion, since words like "phenomena" and "observations," just as "attributes" and "measurements," are used in a way hardly compatible with common language and practical definition (Bohr 1949, 237).

He then makes a point that brings us right back to the first stage of the inductive process and the way it functions, while avoiding the vague "disturbance" parlance that belongs to the interpretive and rather speculative work on the theory:

As a more appropriate way of expression, I advocated the application of the word *phenomenon* exclusively to refer to the observations obtained under specified circumstances, including an account of the whole experimental arrangement. In such terminology, the observational problem is free of any special intricacy since, in actual experiments, all observations are expressed by unambiguous statements referring, for instance, to the registration of the point at which an electron arrives at a photographic plate (ibid.).[2]

In fact, whenever seriously challenged, Bohr reverted to his experimentalist-minded sense of the inductive method. His musings were always on the back burner. Thus, Howard's (2007) impression that the entanglement of the observer and the apparatus is central to Bohr's arguments and crucially defines them is somewhat misleading. Although Howard characterizes one feature of Bohr's views properly, mainly in relation to the Como lecture, Bohr does not develop that point on entanglement as a defining point in his response to the EPR paper, also announced in his Como lecture. Instead, this part of his analysis may belong to a philosophical musing that ought to be recognized as such, and as distinct from the vision of physics that made his breakthroughs possible. After all, his response to the EPR paper is a heterogeneous piece aimed at wider audience. And these musings likely have little to do with the reason why so many physicists have seen him as a clear winner in the debate. Bohr's focus on stating points in the debate that are firmly entrenched in and limited by the details of the key experiments must make a strong impression on many working physicists, and must seem a convincing response to a rather abstract challenge that Einstein was pushing.

15: THE MATURE BOHR AND THE RISE OF SLICK THEORY AND THEORETICIANS

With particle physics, the next chapter, the post-Bohr era begins.
—Abraham Pais (1968, 220).

As I have noted, many current criticisms of Bohr offered by philosophers might convey the impression of a mumbling and tedious middle-aged physicist who projected his authority on young physicists while distracting them with nonnegotiable but somewhat obscure philosophical views. Indeed, many commentators have focused on writings from Bohr's later years, and those were certainly framed more abstractly than his earlier work. In contrast, I have offered an understanding of Bohr that adheres closely to his practical work in the period when he was winning his major fame among physicists and establishing his central mediating place in the community, following the publication of his atomic model when he was a twenty-eight-year-old in 1913. Further, I have shown how Bohr's practical methods and approach changed little in the next decade and a half, when he was producing his complementarity principle, among other things.

It is not unusual for a successful scientist like Bohr to test his ideas and capabilities outside his strict zone of expertise and to write for a nonspecialist audience, once his peers in the field have recognized his contributions as major. As a matter of historical analysis, these attempts should be treated with care and judged on their own, not hastily conflated with previous major contributions and the approaches that led to them; and they especially should not be portrayed as pivotal in Bohr's work.

Discussions of Bohr's philosophical obscurity are too often confined to his later writings. This is certainly true of the perception of complementarity. As Patricia Kauark-Leite (2017, 68) notes, Bohr's recollection of his debate with Einstein, published in 1949, "marks the turning point in the way people conceive the complementarity principle, restricting it to wave-particle duality." As a result, understanding the methodological origins of the principle was almost inevitably sidelined, with prominence given to subsequent and rather assorted philosophical reflections on it. Stefano Osnaghi reflects this widespread attitude when, in connection to the understanding of the notion of the completeness of quantum mechanics that Bohr debated with Einstein, he writes, "The expectation is that Bohr's ambiguities can be partly explained, if not removed, by retracing the dialectical process that resulted in the mature formulation of his argument for the completeness of quantum mechanics" (Osnaghi 2017, 156). Similarly, Arkady Plotnitsky (2017, 180) states that "complementarity, and the non-realist philosophy that accompanies it," were brought to their "radical limit in Bohr's later thinking, via concepts of phenomenon and atomicity, which supplemented his concept of complementarity."

This assessment of Bohr's alleged radicalization of his own views may well be adequate; but if it happened at all, it certainly happened long after Bohr and his collaborators worked out the provisional account, central to the development of the theory, by inducing it from the experimental context. The potential danger of focusing on Bohr's afterthoughts in our critical interpretations and treating them as the most developed version of what was going on earlier in Bohr's work is that of making the assumption that the older Bohr was intellectually more mature and had a privileged vantage point over the younger Bohr who had produced the breakthrough results. This may or may not be true, so the analysis of his intellectual interests and his approach in one period of his work should be carefully disentangled from or connected with the analysis of his work in other periods—especially his actual practice of physics during his breakthrough period—because humans in general, including Bohr, are prone to change their views and aims, often radically.[1]

It is enticing to pin an interpretation of Bohr's philosophical views to the mature Bohr, especially his 1958 ruminations "Quantum Physics and Philosophy: Causality and Complementarity" and "On Atoms and Human Knowledge." Authors often turn to this work when discussing Bohr's general understanding of physical systems and his broader philosophical views. Although nothing in this work seems to straightforwardly contradict Bohr's earlier comments, the statements in it are not situated in a particular experimental context, nor do they seem gradually induced from it. Instead, they are the product of a fairly loose thinking process. This work was written decades after Bohr achieved his breakthrough results in a very different theoretical context, so we should read it with caution and certainly not as a strict guideline for understanding Bohr's methodological approach up to the end of the 1920s. We should rather see it as a clue for interpreting his mature thoughts. Moreover, as I have already pointed out, we need to take into account precisely whom Bohr was targeting. His statements and arguments on the nature of physics in the papers aimed at a quantum physics audience are carefully and directly drawn from his actual practice. His writings on more general topics usually target a wider audience, and do not aim to reflect on the actual practice of quantum physics as directly, or at all.

A Kantian streak in Bohr's approach to physics, and in Bohr's intellectual obsession with the general limitations of expressing physical reality in language (Bitbol 2017; Katsumori 2011; Faye 1991; Murdoch 1987; MacKinnon 1985; Bub 1977; Stapp 1972; Petersen 1968; Rozental 1968, 107), was beginning to be shaped as a somewhat developed general account only when the basic experimental inductive process was winding down. Yet many authors take it as the starting point when they interpret Bohr's earlier work. It is expected that the thought of a scientist of Bohr's intellectual vigor and education will be multilayered, but it is essential to discern the actual nature, place, and role of those layers.

Even Howard's exceptionally insightful analysis (Howard 1979, 2004) is much more in line with the mature Bohr than with the vigor of his work up to the end of the 1920s. Howard emphasizes that Bohr's complementarity approach stemmed from his insistence on the entanglement of the experimental apparatus and the observer. But this is only part of the motivation, a final step perhaps of the emerging account. The step was formulated in explicit terms as a provision under which the physicist can unambiguously ascribe properties to observed objects, but it can hardly be a starting point for understanding the emergence of the complementarity account and the role it played in the 1920s. Complementarity was a product of a much more comprehensive grasp and process, one Bohr gradually developed and adjusted to the experimental and theoretical context over the years, starting with his work on the model of the atom.

There is little to suggest that Bohr's later development of the complementarity principle as a comprehensive account applicable across scientific fields should take precedence over complementarity as a provisional master hypothesis in an orderly inductive process during the 1920s. Bohr's mature ruminations on a general meaning of complementarity often far exceed the limited experimental context in which the concept was initially devised. His thinking about its wider applicability in understandings of causality, biology, or psychology is interesting in its own right (Rosenfeld in Rozental 1968), yet these later developments are often afterthoughts of sorts, clearly far removed from the experimental work to which he clung when devising all three major contributions. We at least ought to be very careful when devising analysis of the sort just described, as it may produce confusions of its own, making Bohr's crucial work in physics and its accompanying formulations unnecessarily and misleadingly obscure. Unfortunately, this seems already to have happened in much Bohr scholarship over the last few decades. That is why this book combines a comprehensive analysis of his practice of physics, especially during his breakthrough period, with his reflections on his practice as they developed over time. Interpretations of the latter, especially in the later period of his career, must be read against the former.

Also, Bohr's attempts to develop his obsession with the limitations of language, which some authors take as a key feature of the obscurantist nature of his philosophical views, never really amounted to a systematic research program if assessed by the criteria of recent academic philosophy. As I have shown here, we can understand them properly only in the comprehensive context of his practice of physics. His considerations of the nature of scientific observation and on the observation of physical phenomena accompanied his work in physics in the form of occasional reflections. They "gradually emerged from Bohr's patient and painstaking analysis of the implications of quantum mechanics" (Rosenfeld in Rozen 1968, 124). Bohr repeatedly reflected on the difference between observations of classical and quantum physical systems. The correspondence principle led him to think of a classical observation in which one's perception is confined as a result of observing a very large number of quanta. And an observation of an individual quantum system could not be performed in the same manner even in principle, as Heisenberg's famous thought experiment with the microscope was intended to illustrate.

As I have shown previously, Bohr kept these instructive general reflections to the margins of his major works—for example, in his aforementioned paper "Atomic Theory and Mechanics," published in *Nature* in 1925, which anticipated the basics of the complementarity account. If his general reflections could clearly lead to progress in understanding the experimental context at stake, he carefully and gradually included them in the finished works. The reflections were tools meant for the consideration of conceptual issues and new directions in which to develop the theory, working on the margins of the main inductive construction.

Bohr never questioned the key adherence to the experimental context when working on what turned out to be his major breakthroughs. It was the anchor of his entire approach to physics. Thus, the main tool in his conceptual consideration of a "thought experiment" was an outline or even a detailed drawing of the experimental apparatus. It was supposed to demonstrate what exactly could be observed in the first stage of the inductive process, and thus to set clearly the conditions for the elicitation of a more general hypothesis. The contrast with Einstein's more abstract thought experiments and the aim and vision of their important aspects in the debate is striking. In assessing and responding to Einstein's "black box" thought experiment questioning the indeterminacy relation between time and energy, as well as a series of similar experiments Einstein proposed to him over the years, Bohr worked out his own thought experiments that worked rather differently:

> Even if these experiments existed only in thought and never should nor could be carried out in practice, it was important to stress the considerable mass of the individual parts of the apparatus and their firm connection with each other. The apparatus was therefore drawn a second time in such a way that it visualized the thickness of the plates, the massive screws which held the parts in place on a heavy, solid base, the structure of clocks with cog-wheels etc. (Rozental 1968, 180).

Thus there is very little if any evidence that Bohr initially developed the complementarity principle as a comprehensive philosophical view and then applied it to a particular experimental context. Rather, on the basis of everything discussed in this book, this most abstract and general aspect of the complementarity principle, which developed from the principle as a provisional account of a particular experimental context, was a product of the larger philosophical concerns Bohr displayed even as a very young man. Once the crucial inductions were accomplished and a novel master hypothesis was established, he slowly addressed these more abstract concerns, attempting to develop a more general idea of complementarity and to apply it to the understanding of life in biology, to a scientific approach to cognition, and even to anthropological research. The fruitfulness of these attempts is open to debate, but they are certainly very different from his work on inducing quantum mechanics. They are, in fact, much like Schrödinger's fascination with and study of Indian and ancient Greek philosophy, especially Schrödinger's understanding of the principle of continuity in that context, compared to his development of wave equation and his wave interpretation of microphysical quantum phenomena. Bohr's central role in the physics community brought much light to bear on his subsequent philosophical work aimed at a general audience, and opened the door to more obfuscations, in contrast to Schrödinger's case. Many scholars have interpreted this aspect of Bohr's work as an ideological matrix for understanding his approach to his major contributions—a rather naive interpretation from the likes of which Schrödinger was luckily spared. And Bohr's later explicit reflections on philosophy and quantum mechanics, published in the 1950s, may be as relevant or irrelevant as Heisenberg's (Heisenberg and Bond 1959) recollection of his own philosophical views in the 1920s.

*

I cannot understand what it means to call a theory beautiful if it is not true.
—Niels Bohr (in Rosenfeld 1968, 117)

This vigorous dismissal of beauty as a criterion for assessing a physical theory—even one that may be incorrect, or a perhaps speculative and unconfirmed creation, is a mark of a staunchly experimentally minded physicist. And the dismissal is in stark contrast to the role played by the concept of symmetry in modern physics. Ironically, soon after Bohr pronounced the words in the quotation above, physical phenomena themselves turned out to be defined by essentially basic aesthetic relations as the core posits of physical theory. A vision of physics relying on deep theorizing and complex mathematical devices had already been anticipated and grounded by Emmy Noether's groundbreaking theorem, which demonstrated the equivalence of the laws of conservation and a set of basic physical symmetries. This vision of physics flourished through the rest of the twentieth century and the first decades of the twenty-first. It is a fair assessment, then, that "Bohr created atomic physics and put his stamp on nuclear physics," but also that "with particle physics, the next chapter, the post-Bohr era, begins" (Pais 1968, 220).

During the periods of major breakthrough discoveries in physics, the phase of the experimental grind is typically followed by a phase of intense focus on polishing theoretical accounts of relevant phenomena. These subsequent theoretical efforts may become significantly removed from the actual experimental context. Recent developments in string theory and in algebraic approaches to quantum mechanics are good examples of this. The experimental grind aims at providing the basis for formal and theoretical tools that can account for all the relevant experimental results.

The high-level quantum theory, the subject of contemporary discussions, began to emerge with Paul Dirac's insights into symmetries related to quantum phenomena and von Neumann's refurbishing of formalism. These two results marked a phase change in the development of the theory. Before then, it was not possible to devise or assess well defined and theoretically refined distinctions and alternatives. Toward the end of his Como paper, in fact, Bohr anticipated this phase change in the development of quantum mechanics. Moreover, he saw the Institute for Theoretical Physics in Copenhagen as a place to accommodate physicists whose work approached quantum phenomena in diverse ways, including ways he had not pursued himself. After all, this was the reason why he had placed it in close proximity to the Institute of Mathematics. This aspect of his approach to quantum theory and physics in general is too often overlooked; and when it is mentioned, the Copenhagen "school" is often portrayed as sectarian.

Dirac arrived at Bohr's institute in 1926 (Klein 1968, 87), and his work had a strong influence on researchers there. In particular, it had become clear that quantum mechanics was a single theoretical framework, and that future development should aim to enforce that, not to account for experimental phenomena piecemeal. Dirac's development of radiation theory and his account of the electron that captured its nature in terms of the symmetry relations was probably the turning point in the introduction of a high-level theory. The early quantum field theory was another development initiated by Dirac. Pascual Jordan, Eugene P. Wigner, Wolfgang Pauli, and Heisenberg contributed to this work.

From the very beginning, Heisenberg exhibited the urge to "leave the models and take the step over to mathematical abstraction" (Heisenberg 1968, 98). Although the result of his approach could not be generalized, for the reasons we have discussed, it was perhaps the first step in the direction that the theory soon took. And, as we have seen, Heisenberg's theoretical zeal was justifiably tamed by Bohr in light of relevant experiments. Heisenberg later admitted: "To what extent it would be possible to use these mathematical methods to build a complete theory was, however, still uncertain at the time" (Heisenberg 1968, 100). At the same time he said, "Mathematical schemes had for me a magical attraction and I was fascinated by the thought that perhaps here could be seen the first threads of an enormous net of deep-set relations" (ibid.). This was the kind of motivation that inspired Heisenberg, Dirac, Hermann Weyl, and others to start constructing the quantum field theory.

Bohr was suspicious of Heisenberg's and Pauli's attempts to develop quantum electrodynamics—that is, to try to apply quantum mechanics to electromagnetism. As Rosenfeld commented, "He always regarded with deep suspicion any theory not solidly anchored in some concrete reality . . . and the difficulties encountered in the attempted extension of quantum theory to electrodynamics seemed to him so remote from any familiar physical situation that he was not easily persuaded to take them seriously" (Rosenfeld 1968, 114–15). This sort of reaction accorded with the vision of physics that had led him to his major breakthroughs. Heisenberg, who nurtured a much more mathematically minded, as it were, approach to physical phenomena than Bohr, stated that "Bohr was primarily a philosopher," in that he was working out his account intuitively through conceptual assessments. Perhaps in a somewhat exaggerated manner, Heisenberg points out that "mathematical clarity had in itself no virtue for Bohr. He feared that the formal mathematical structure would obscure the physical core of the problem, and in any case he was convinced that a complete physical explanation should absolutely precede the mathematical formulation" (Heisenberg 1968, 95–98).

Bohr certainly was not a physicist who immediately turned to mathematical apparatus to dissect the phenomena at stake. Heisenberg said, "He understood that natural philosophy in our day and age carries weight only if its every detail can be subjected to the inexorable test of experiment" (Heisenberg 1968, 95). This attitude went along with a sense of "the danger of extrapolating formal procedures beyond the domain of experience for the description of which they had originally been devised, and the necessity of seeking a solid foundation for our lofty abstractions in some simple concrete aspect of phenomena immediately accessible to observation" (ibid.)—something crucial for balancing the complex development of the theory and tying all the loose ends together, even in a provisional manner. This is particularly apparent in this treatment of new formalisms, as discussed in chapter 11. Yet, as we have seen, Bohr was fully aware of the start of the math-driven phase in the theoretical development of quantum mechanics. His worry only concerned whether particular pursuits such as quantum electrodynamics were adequately embedded in the experimental work.

It is also possible that an alternative theory such as Bohm's would have seemed out of place in the mid-1920s, since it was not even clear whether any of the formalisms were there to stay. De Broglie's arguments played a role but were very far from such a theoretical framework, because both theory and experiments were in constant flux. Counterfactual historical scenarios of this sort that, had they happened, may have speeded up the development of physics (Becker 2018; Maudlin 2018) may not be plausible given the actual historical context. In any case, Bohr's approach was reasonable, as was that of the quantum community around him, and they both gradually achieved the aim they set out to accomplish. If other historical trajectories of building quantum mechanics were possible, including those based on what today many would consider brilliant theoretical anticipations, the trajectory that did occur was nonetheless the result of a careful and reasonable experimentally led strategy.

How long each phase of a theory's development will take depends on the complexity of the phenomena and the complexity of the experimental and theoretical tools developed to account for them. The premature insistence on theoretical slickness can impede efforts to fully and comprehensively account for all the available evidence and formulate adequate basic theoretical tools that can be further refined and developed. As I have argued, this was an essential feature of Bohr's argument with Schrödinger. How fruitful the phase of deep mathematical theorizing has been in modern fundamental physics is a matter of further analysis (Hossenfelder 2018).

Constant direct experimental triangulation and the flux of theory and experimentation of the sort upon which Bohr insisted are not achievable in contemporary high energy physics laboratories. The experiments do not offer a constant stream of variety of evidence matching that in physics labs of the early twentieth century. Producing even one line of evidence takes much longer, often decades; so the theory has typically been ahead of the experimentation. In such an experimental context, the inductive process—the shaping of theoretical concepts by the experimental particulars, which was so crucial for the emergence of quantum mechanics and Bohr's approach—has gradually been sidelined. This has not happened all at once but in several phases, with the emergence of ever-larger laboratories and ever-longer-running experiments. Bohr's vision of physics as a constant and intensive cooperation in the community warmed him to the idea of the European Council for Nuclear Research (CERN), a pan-European laboratory for particle physics. He was actively involved in its creation, and was asked to open the large proton accelerator (Rozental 1968, 186). Ironically, this development spells the end of the kind of physics he nurtured.

ACKNOWLEDGMENTS

My interest in the topic of this book was initially motivated by reading Thomas Kuhn's book *Black-Body Theory and the Quantum Discontinuity, 1894–1912* in a graduate course directed by Jagdish Hattiangadi at York University in Toronto some twenty years ago. Jagdish and I started a long conversation on the history of quantum mechanics from which I benefited immensely. Historical works by Helge Kragh and insights he generously offered in our conversations were invaluable, as was Allan Franklin's and John D. Norton's encouragement and the time they took to read and comment on the ideas that eventually made it into this book. Milan Ćirković was the only person who read the very first draft of the book and offered friendly but precise and knowledgeable comments and advice. Anonymous referees with the University of Chicago Press were exceptionally patient and helpful, as was the editor Karen Merikangas Darling, who navigated me through the process with great tact and skill. Leon Kojen offered great advice in terms of the structure of the arguments. I am also grateful to Andrei Khrennikov, Don Howard, Christian Camilleri, Christian Joas, Lidija Živković, and Živan Lazović for our exchanges of ideas; to Elizabeth Thompson for proofreading parts of the text; to Renaldo Migaldi for immaculately editing the manuscript; and to Eric Winsberg for great advice in the key stages of the publishing process. I am also thankful to many graduate and undergraduate students at the University of Belgrade and the University of Pittsburgh who commented on some of my ideas. My wife and colleague philosopher, Ljiljana Radenović, encouraged me more than anybody else, and offered crucial advice during the long writing and publishing process, while my parents provided a firm foundation for this endeavor. I hope that those whom I have not thanked here know how much I am grateful for their help.

NOTES

INTRODUCTION

1. I noted some of these points concerning this puzzle in Perovic 2013.

2. For a comprehensive criticism of Lakatos's view, see Radder 1982 and Hettema 1995.

3. "Bohr and Heisenberg's approach to quantum physics—known as the Copenhagen interpretation"—a rather loose and historically unjustified characterization of all three—"was pervaded by the same vagueness that Bell had found in his quantum physics courses" (Becker 2018, 17), while the "obscurity of [Bohr's] thought merely added to [his] sagelike qualities" (Becker 2018, 32).

4. "There is this horror of the way Niels Bohr ruled things. . . . [Niels Bohr is] one of the people [I would most want to meet], if I could pick to meet a historical figure. He must have been the most . . . charismatic human being in the history of the world. There was just this long string of brilliant people who would spend an hour with Bohr, and their entire lives would be changed. And one of the ways in which their lives were changed is that they were spouting gibberish that was completely beneath them, about the foundations of quantum mechanics, for the rest of their lives. And you want to know: How did this guy do this? . . . Boy do I want to meet this guy." https://www.preposterousuniverse.com.

5. For a comprehensive list and analysis of such influences, see Heilbron 2013 and Aaserud and Heilbron 2013.

6. For recent examples, see Dieks 2017; Zinkernagel 2016; or De Gregorio 2014. We will encounter more work of that sort as the book progresses.

7. A sketch of my idea how Bohr's method led to his model of the atom, and not only to his complementarity principle can be found in Perović 2019.

8. I am particularly indebted for thinking in this direction about Bacon's understanding of induction, and about induction more generally, to insights in Hattiangadi 2006; Sargent 2001; and McMullin 1990, 1970.

9. See Kragh 2012, 367.

10. It will become clear that critics such as Maudlin miss the target when accusing Bohr of deliberately giving up on the ontology and dynamics of physical theory that should meet their stringent standards of clarity. Maudlin writes, "Not every physical theory makes any pretense to provide a precisely characterized fundamental ontology. A physical theory may be put forward with the explicit warning that it is merely an approximation, that what it presents without

further analysis is, nonetheless, derivative, and emerges from some deeper theory that we do not have at hand" (Maudlin 2019, 5). In fact, such warnings abound in Bohr's work all the way to the 1930s. The theories he was creating were explicitly provisional; and at that historical moment it is hard to see how they could have been any different, or how pushing "ontological clarity" too far could not fail to be counterproductive.

11. See, for example, the papers by Giere (2010), Van Fraassen (2008), de Chadarevian and Hopwood (2004), and Morgan and Morrison (1999).

12. If we wish to put a label on him, with all its limitations, Bohr was an experiment-oriented practicing inductivist.

13. Becker's (2018, 17) attitude nicely illustrates a frequently used, semi-popular, simplistic version of Bohr's view on this key point: "The godfather of quantum mechanics, Niels Bohr, talked about a division between the world of big objects, where classical Newtonian physics ruled, and small objects, where quantum physics reigned. But Bohr was maddeningly unclear about the location of the boundary between the worlds."

14. See Schickore 2011 for a detailed review of the different accounts of the relationship.

15. Achinstein (1974) developed this argument in the 1970s; Schickore (2011) and Burian (2001) advocated it more recently, and Hasok Chang has practiced it in his work. Giere (1973; 2011) criticized this view, arguing that philosophical analysis of science must confront facts and build theories and models in much the way science does. Laudan (1977) defended a similar view. I agree that this sort of analysis has value—I studied high energy physics following such a model of inquiry—but it is certainly not the only valid approach to the philosophically motivated study of science.

16. This approach contrasts with the tradition of "historicism," and especially the Hiedeggerian notion of interpretation as discussed by Shickore. In any case, the reader should not expect the process of developing a cogent and compelling account to be immediately transparent, just as she should not expect to understand fully the method that produces a scientific paper just by reading it (Schickore 2011, 472–73). She ought to judge the work based on its cogency and on whether she finds it compelling.

17. See Schindler and Scholl (forthcoming) for the account of case studies in philosophy of science as analogous to the use of model organisms in biology.

18. The same goes for a variety of views on what exact role hypotheses play in experimentation and what exactly constitutes experiments (Franklin and Perovic 1998). To what extent the experimental practice and reflections on it are widespread in science, and in what contexts, is another matter.

CHAPTER 3

1. A crucial ambiguity concerning this point in the work of Camilleri (2017) and Camilleri and Schlosshauer (2015) is that they label Bohr's attitude to the

experiments that produced his view of classical properties as "functional epistemology," although they themselves point out correctly that Bohr shied away from ready-made philosophical doctrines, including epistemological ones. Moreover, they say that Bohr's treatment of classical states is a result of his role as a philosopher of experimentation. This has the ring of a somewhat strained and potentially misleading view. We need to remove ourselves far enough from the enticing categories of the contemporary philosophy of science community to fully understand Bohr's work, even though using such categories may make our analysis more appealing. Bohr's account of classical states is but a feature, albeit an important one, of his general approach to physics—not a product of a stand-alone philosophical pursuit of a curious philosopher of experimentation.

2. Dorato (2017, 134–35) correctly acknowledges the distinction between sharply divided classical and quantum notions and descriptions on the one hand and the entanglement of the subject and instrument on the other. The latter, he says, is certainly distinct from the former, with a line dividing experimental work and hypotheses from more abstract and general work and more general and abstract hypotheses. Yet he believes that Bohr's motivation for the former was philosophical, that is, that of a realist, along with "the intersubjective validity of physical knowledge" (ibid., 135) and five related reasons. My account of how Bohr built his major contributions will demonstrate that this sort of philosophical consideration was not high on Bohr's priority list of methodological and epistemological concerns; rather, it was only one of many considerations in the last stages of building the master hypothesis (more especially with respect to his complementarity account than his model of the atom).

3. See also Schaffer 1989; Klaus 1997; Beller 1999; and Harré 2003.

4. The aim of the discussion in this section is to philosophically clarify Bohr's view of classical and quantum properties, not to provide a historiography-based analysis of its particular aspects, parts of which can be found elsewhere (Camilleri 2017; Camilleri and Schlossauer 2015).

5. In other words, we use the notion of hypothesis as a catch-all term encompassing a hierarchy sorted out and based on the proximity to the experimental setup. Francis Bacon distinguishes the notion of the hypothesis from that of the axiom to emphasize that axioms are drawn directly from experimental results, whereas hypotheses result from speculation that's only superficially adjusted using biased and hand-picked experimental results. The modern use of the notion of axiom belongs to the context of mathematics. The notion of hypothesis, however, refers to numerous and sometimes quite various things. Bacon's axioms are precisely what we now consider hypotheses arrived at by experimentation or, at the higher level of abstraction more fittingly, as theoretical postulates. The latter term was often used by twentieth-century physicists to explain the theoretical propositions they elicited from experiments.

6. For a variety of models and their uses by Thomson, see Heilbron 2013, 48; and for models in general in that period, see Kragh 2102 and Seth 2010.

7. I turn to this case in more detail in the next section.

8. Discovery of the anomalous Zeeman effect and Stark's effect are part of the same process of including a lower hypothesis in the theory, as well as other relevant experiments I will discuss in due course.

9. It is correct to say that "Bohr claimed that classical concepts are also necessary to *interpret* the information carried by the pointer of the apparatus" (Bächtold 2017, 241); but long before the *interpretation* as a higher-level theoretical work to which the author refers, there is an *interpretation* of experimental particulars in producing lower-level hypotheses. This is what Bohr refers to first and foremost when introducing the concept of the classical. The reasons for making a classical/quantum cut are not philosophical and interpretive, but have to do with Bohr's understanding—stemming from his own practice of physics and the way it was practiced in the community at the time—of the divide between the experimental process and subsequent theoretical work.

10. See Aaserud and Heilbron 2013, 112–15.

11. See also Dickson 2002 and Bokulich and Bokulich 2005.

12. Howard's (1979) analysis points out the importance of the role of the observer-object entanglement in Bohr's theoretical stance. Yet it falls short of recognizing the meaning and importance of this separation; the entanglement thesis concerns how an already established "classical" record of observational particulars (results of stage 1) can be *interpreted*—i.e., it concerns the nature of the hypothesis elicited by the results, without commitment to a lack of ambiguity.

13. This is why his account of induction is akin to that of Bacon, perhaps more than to the work of any author in the same tradition (Perovic 2013). This may not be surprising; despite working in different periods, their accounts stemmed directly from their scientific practice, which led them to emphasize the same point as central.

CHAPTER 4

1. See Casimir 1968, 110.

2. The repetition of an experiment is merely one concern in the induction of hypotheses, and it is not the most important one, contrary to what those empiricists interested in the Humean problem of induction might think. The repetitious aspect of experimentation is necessarily a part of a thorough experimental work, as we cannot fully rely on our own senses or on a single individual experimental run in which they are used. This aspect takes a particular and more complex shape in modern particle physics, even in the early stages: the repetition is typically always part of subsequent experiments performed on another apparatus, usually varying the experimental context to probe previous results and provide robustness (Franklin and Howson 1984).

3. See also the work by Gooding (1990) and Pickering (1984), and a debate between van Fraassen (2009) and Chalmers (2011). For the role of multiple de-

terminations of a phenomenon through a variety of experimental designs, see Woodward 2006; Stegenga 2009; and Franklin and Howson 1984.

4. Duhem (1954) made this simple point, and philosophers like Kuhn (1970) and Feyerabend (1970) turned it into a forceful but rather crude argument for the radical theory-ladenness of evidence. The experimental process is thus biased by such operational theories in a certain direction. Yet operational theories can be questioned if needed, if the experimenter is vigilant enough and focuses on the observational content more than she conforms to the expectations of the operational theory itself. When the experimenter realizes that the operational theory is biased, she can test the phenomenon with an apparatus that produces data based on a different operational theory. To use an example given by Franklin (1984), in the case of measuring dependence between the rise in temperature and the expansion of substances, the mercury-based thermometer will not suffice, as it already presupposes that the rise in temperature is measured by the expansion of the substance. The experimenter can simply use the constant-volume gas thermometer instead, avoiding such a bias, or even calibrate her mercury-based thermometer with it. In fact, the complex history of measuring and trying to understand temperature (Chang 2004) is little more than a history of experimenters' struggles to avoid such biases of operational theories and so is the history of quantum mechanics.

5. This sort of bias in the discovery that beta decay is the manifestation of a novel force was overcome by the use of diverse techniques. Researchers discovered that beta-spectrum is continuous only when there is "qualitative guidance" from other phenomena (Pais 1986, 144). In addition, the question of how many electrons are emitted in the decay was answered only after the introduction of the cloud chamber (ibid., 145). The initial question addressed experimentally by Otto Hahn and Lise Meitner was how electrons are absorbed when the ray passes through matter in beta decay. As electrons scatter in all directions (as detected by incoming rays hitting metal foils), researchers did not understand what beta rays could be. However, by ingeniously building on previous techniques of absorption in 1908, Wilson discovered the beam that is prepared as monoenergetic does not follow the exponential law of absorption that was accepted as a general law at the time; he therefore concluded that the rays lose energy *gradually* when traversing. They are absorbed exponentially only when traversing materials of different thickness, and the rays must be heterogeneous in such cases because homogenous rays do not absorb exponentially.

Wilson constructed an apparatus in which, after going through a slit, the rays entered a homogenous magnetic field perpendicular to their velocity bands, which set them in circular motion. They ended up making a semicircle as they entered another slit and hit the metal foil of variable thickness. The rays going through the slit (absorbed by the metal foil) had homogenous velocities; the current of the rays traversing the foil was measured with the electroscope. The correct conclusion, based on the observed currents (and for each tested

velocity), is that beta rays recede gradually; that is, they do not absorb exponentially. Wilson's multifaceted experimental apparatus became a classic in
absorption experiments. Yet, as insights in his experiments stemmed from a
very specific question, he did not go on to prove the continuous nature of the
spectrum (ibid., 153). It was thus assumed that the electrons were discrete.
Challenging this assumption required two very different techniques that
probed two aspects of the investigated phenomenon. Experiments with the
photographic plates that recorded beta decay seemed to indicate the separation
of spectral lines. Therefore, Hahn and Von Bayer applied Wilson's magnetic
separation of electrons of different velocities using the photographic detecting
technique, even though this detecting technique was in its initial stage of development and was not well understood. After the electrons of different velocities
looped through the semicircular orbit, they hit the photographic plate; the
blackened parts on the plate were supposed to correspond to the initial velocity
spectrum. After further experiments and a thorough rethinking of their initial
position, they concluded that they were recording the dispersion of beta rays,
not their absorption (which assumed a separate chemical substance for each
beta line as the substances differed in absorption characteristics). The differences in blackening were caused by the beta rays' loss of intensity on the way to
the plate. This was a subtle step on the path to understanding nuclear spectra
through photographic detection.

Similarly, a number of experiments—starting with Hughes, Richardson, and
Compton in 1912 and Millikan in 1916—addressed the issue of the photoelectric effect, each inventing new detecting techniques and/or new experimental
apparatus that involved sources of radiation and its control. A subsequent
experiment that led to the famous Compton scattering discovery had already
been performed on a versatile apparatus that combined the techniques of these
previously mentioned experiments.

6. A theory does not always implicate a bias. Sometimes a particular theory
can motivate the experimenter to challenge the accepted bias. For example,
Friedrich Paschen formulated his spectroscopic selection rules with the help of
Sommerfeld's theory (Kragh 1985). This sort of help in selecting experimental
particulars was fairly common in spectroscopy at the time.

7. Sommerfeld's theory of diffraction, which he developed in his habilitation thesis and presented in a distilled form at the Solvay conference, was
probably a very important factor in Thomson's development of the theory (Seth
2010, 147).

8. The 1890s debate on the nature of radiation was addressed in two experimental centers and played a decisive role in constructing new radiation laws as
the initial phase of the development of quantum theory. Radiation and electric
radiation were probed by Hertz in his laboratory. In the debate, Pringsheim,
Paschen, Helmholtz, and Julius asked whether newly discovered line spectra stemmed from differences in chemical or physical properties. The newly
founded Physikalisch-Technische Reihstanstalt in Berlin, headed by H. Rubens

and E. Fox Nichols, used very different techniques and background theory than Paschen. The former studied radiation by measuring units of light, while the latter measured normal spectra bolometrically. The debate culminated in 1899 over whether short or long wavelengths were appropriate subjects of study (Kangro 1976, ch. 7, 147–52). The disagreement on the appropriate energy range to be studied was essential to the treatment of the issue, as it resulted in construction of two different apparatus and, accordingly, two operational hypotheses that ultimately were unified in quantum theory: the photometry of the light source, whose temperature was supposed to be determined photometrically through the analysis of different heated materials, and the bolometric measurements that relied on measuring electromagnetic radiation as it depended on the changes in electrical resistance.

9. This change in the key concept was reflected in the design of subsequent apparatus as well. Zeeman's use of grating in the experiments with spectral lines was based on Lorentz's theory of the electron.

10. Stegenga (2009), and Franklin and Howson (1984) develop arguments on the difference between replicating and varying experimental setups.

11. Another episode of experimentation instrumental for the development of quantum theory made the importance of the intricacies in the selection process of experimental particulars transparent. In his attempts to crack its spectrum, Wilson's magnetic separation technique was united with the detection by counters, a detecting technique very different from the photographic plate used previously. The importance of this novel technique appeared very quickly. Initially, Chadwik complained: "I get photographs [of seeming lines of beta spectrum] easily, but with the counter I can't even find the ghost of the line. There is probably some silly mistake here" (Chadwik in Pais 1986, 159). But this was not a silly mistake, as Chadwik soon realized. Rather, the counters recorded only the lines of beta decay after the electrons of different velocities (by varying magnetic field strength) looped through the semicircle and discharged in the electric potential (the actual counter) between a fine needle and a metal plate. The obvious conclusion was that the spectrum of beta radiation was continuous. As Hahn soon understood, the photographic plate recorded a dispersion that produced fake lines if developed in a particular way.

12. Schrödinger's wave-mechanical interpretation of microphysical quantum states and process in the 1920s violated this requirement; it put a metaphysical concern, namely the principle of physical continuity and intuitions based on it, ahead of the adequate grasp of experimentally drawn hypotheses.

13. It would be strange to hold Bohr accountable to higher standards of modeling than we do current physicists. The literature on modeling insists on ad hoc and inaccurate nature of models as heuristic tools that nevertheless perform their key tasks.

14. Later on, in the experiments that were instrumental in devising the complementarity approach to quantum phenomena, the experimenter could interpret hits on the screen recorded in the double-slit experiment either as

marks left by a bunch of particles forming a pattern in the form of a wave front, or as traces of the wave that seem to appear as particle marks. Similarly, a track in a Wilson cloud chamber can be taken either as a pattern of the atom's ionization wave front or as a trace of a particle that whizzes by.

15. Thomson's model did not suffer from the problem of mechanical instability, but Bohr thought Rutherford's model superior on other counts.

16. See Bohr's letter to Rutherford dated 6 July 1912 (Bohr 1972–2008, vol. 2, 577).

17. For the distinction between the mechanical and radiative instability of the nuclear atom and their respective importance for Bohr's model, see Heilbron and Kuhn 1969, 241n81.

18. See also Heilbron 2013, 26–27.

19. For a detailed account of deriving his model by drawing an analogy with Planck's law, see Helibron 2013, 30–32.

20. See Heilbron and Kuhn 1969, 216–21, for a detailed analysis. The authors summarize Bohr's results and its importance in the following way: "At the very least the difficulty with magnetism strengthened and confirmed Bohr's conviction that the usual mechanical laws broke down when applied to rapidly moving electrons; and even more than the radiation problem, it isolated the breakdown in the behavior of electrons bound into atoms" (ibid., 222). The problem with magnetism "focused his attention on the question of bound electrons, which would ultimately become for him, the problem of atomic structure" (ibid., 223).

21. It was a novel consideration of quantum of action independent of the quantization of energy and the electrodynamics mechanism (Seth 2010, 158). Seth argues that Planck set out the problem as statistical, while Sommerfeld took a dynamic approach and insisted that such bold dynamic hypotheses, as opposed to mere statistical models, were crucial for the advance of physics (ibid., ch. 5).

22. See relevant details of the overall account in Aaserud and Heilbron (2013, 182).

23. Aaserud and Heilbron (2013, section 2.6) discuss various stages of Bohr's application of Planck's law to account for the rules of distribution of spectral lines by the values the quantized radiation rules.

24. Heilbron (2013, 20) explains in detail why Bohr made a particular move, such as taking Hanson's remarks seriously.

25. See Kragh 2012, 365.

CHAPTER 5

1. Such hypotheses seem to correspond to those Bacon labeled imperfect axioms: "Imperfect axioms as they occur to us in the course of the inquiry . . . are . . . useful if not altogether true" (Bacon 1874, vol. 5, 136).

2. For a detailed analysis, see Kragh 1985, 84.

3. Moreover, the method of deriving any higher hypotheses based on these experimental hypotheses differed substantially and depended on how these patterns should be deciphered. On the one hand, Conway and Ritz grouped the lines into novel but modest lower hypotheses, and these served as the basis for Balmer's and Rydberg's more substantial generalizations (connecting frequencies of electrons with the spectral lines, as opposed to wavelengths of vibrating atoms prior to that). These in turn were the key to Bohr's quantum model of the atom. After 1919, on the other hand, Sommerfeld skipped intermediary hypotheses altogether, drawing "half-empirical" inferences from a lower hypothesis on the distribution of the lines.

4. See Aaserud and Heilbron 2013, 186–89.

CHAPTER 6

1. We should bear in mind that, though Bohr and Rutherford were both very careful and skillful in drawing hypotheses from the experimental context, the inference of a bold general hypothesis was something that the more conservative Rutherford was less likely to undertake. Rosenfeld (Rozental 1968, 46) suggests that it was perhaps a matter of intellectual background, and that the difference between doing physics in Britain and doing it on the continent pushed Bohr to pursue the inductive method to its conclusion, however tenuous it might be.

2. Bitbol (2017, 48) writes: "The patchwork structure of Bohr's model of the atom does not necessarily entail inconsistency, provided the pieces of this patchwork are ascribed a precise but limited use in certain restricted theoretical contexts." Yet he argues that the model lacked a theoretical unity, supposedly a widespread ideal in physics, which was in fact foundational only for some physicists at the time.

3. See figure 3 in Kragh 2013.

4. See Casimir's detailed description of Bohr's account of the one-way optical system proposed by Raleigh (Casimir 1968, 112). It shows the method of devising a general hypothesis from experimental conditions in a nonformal yet clear manner.

CHAPTER 7

1. In fact, forty years passed before the basic hypothesis of the Standard Model of particle physics concerning the existence of the Higgs scalar field responsible for masses of elementary particles was tested in the Large Hadron Collider at CERN. This is a remarkably different experimental context than the one in which Bohr's approach flourished.

2. Hypotheses are a notion we have defined.

3. John D. Norton's (2003) concept of material induction seems a more appropriate general account for characterizing the coordination of complex evidence into a general hypothesis.

4. For example, Weinert's (2001) account of Bohr's induction of his model of

the atom deals with its intricacies, but it does not offer insight into the nature of a comprehensive adjustment to the overall context that led to the model.

5. In a potentially misleading fashion, Heilbron makes a similar point when stating that Bohr advanced the view of "multiple truths," or various versions of the same truth, in his work. The "truths" are really multiple iterations of a master hypothesis, each one based on more adequate intermediary hypotheses, thus tying it more adequately with multiple lower hypotheses.

6. See Rosenfeld's letter to Bohm (6 December 1966, L. Rosenfeld Papers, Niels Bohr Archive, Copenhagen) and Bohr's interview to *Izvestia* in 1934. See also Camilleri 2017 for further discussion of this point.

PART 3

1. As translated by Martin J. Klein in Prizbram 1967, 31. Also accessible at http://healinggeneration.com/hiddenblog/wp-content/uploads/2010/06/LettersOnWaveMechanics_1928-39.pdf.

CHAPTER 8

1. Some of my earlier ideas and arguments on Bohr's method in relation to his complementarity account and Bacon's induction (Perovic 2013) overlap to an extent with the more thoroughly developed account in this part of the book.

2. See Camilleri 2019.

3. He substituted classical parameters in the analysis of frequencies and intensities of radiation with Fourier expansions.

4. See Duncan and Janssen 2009.

5. In fact, Born was the first to realize that matrices can replace classical position and momentum parameters (Camilleri 2009, 21).

6. This was a precursor to the "shut up and calculate" attitude that allegedly took over post–World War II physics; not Bohr's approach, as Tim Maudlin has repeatedly stated.

7. Translation in Camilleri 2009, 52.

8. See also Wessels 1979, 313.

9. For historical analyses of relevant work, correspondence, notebooks, and the exact theoretical sources on which Schrödinger built his account, see Klein 1964; Raman and Forman 1969; Hanle 1971; Wessels 1979; Kragh 1982; Mechra and Rechenberg 1982; and Joas and Lehner 2009.

10. The q-space is a phase space describing all the possible states of the system including, for example, spatial properties and momentum.

11. Heisenberg, who witnessed the meeting, recollects the details in Heisenberg 1968, 103.

12. See Heisenberg 1968, 103.

13. Bohr, Kramers, and Slater advocated giving up the notion of conservation of energy in transfer of radiation, as it was necessary to bring a model to agree with the experimental context. The length to which they were prepared to

go, and perhaps to overreach, in giving up established presuppositions of basic physical principles to account for the experimental context shows the implausibility of the view that Bohr's physics was principally driven by something other than the bottom-up inductive synthesis of the experimental results.

14. An exhaustive analysis of this episode appears in Stuewer 1975 and Perovic 2006.

15. See Kragh 2012, 200–201, and Jammer 1989, 191.

16. See Bohr's reaction to the idea in 1913 (Bohr 1972–2008, vol. 2., 166–67).

17. Camilleri (2009, 108) points out that Bohr did not discuss mutually exclusive experimental arrangements in his Como lecture, and he takes it up only in the later development of complementarity. This is correct, but expected, as such discussions belong to post hoc theoretical analysis rather than to the formative layer of the hypothesis.

18. Schrödinger tried but failed to assimilate this experimental fact into his account (Schrödinger 1927a, 35; Schrödinger 1927b; Perovic 2006).

CHAPTER 9

1. See, e.g., Schrödinger 1930; von Laue 1934; Cassirer 1956; Schlick 1979; McMullin 1954; Jammer 1974; ; Camilleri 2009; and Hilgevoord and Uffink 2016.

2. These foundational accounts are not necessarily in disagreement with my account, but we will avoid dwelling on that issue here and focus instead on our main goal.

3. See D'Abro 1951, section 30, for an early and elaborate reconstruction of this particular explication of the uncertainty principle in the 1928 *Nature* paper.

4. He also spelled out his views at the fifth Solvay conference that same year.

CHAPTER 10

1. To use another example, Einstein's attempt to explain the burning problems of electrodynamics was successful precisely because he took into account, and then synthesized in an adequate hypothesis, a number of phenomena as they were manifested in relevant experiments and as they were explained by hypothesis within particular limited domains (Norton 2014). This approach was unlike partial attempts that tried to force the phenomenon into an existing theoretical framework of classical electrodynamics without attempting a wider grasp of the experimental context. The latter is a legitimate and sometimes useful approach, but certainly not a formula for a successful inductive process within a complex experimental context in search of an appropriate hypothesis to unify it.

2. Such an ascent is the only one that can remove the "mask" from natural objects. In contrast to biased anticipations, "interpretations . . . are gathered piece by piece from things which are quite various and widely scattered, and cannot suddenly strike the intellect" (Bacon 2000, 28, xxviii).

3. See Becker 2018 for the most recent counterfactual historical consideration of quantum mechanics of a similar sort, focused on Bohm's view.

4. For the relevance of the relativistic wave equation that Schrödinger developed and Bohr commented on in this debate, see Joas and Lehner 2009, 349.

5. See Dieks 2017, 304–5, for a more elaborate argument of Bohr's understanding of "individuality" in wave mechanics.

6. See Perović 2017 for a somewhat more narrow attempt of this sort.

7. For recent work on the issue, see Sparenberg and Gaspard 2018.

8. For an account of the similarities and differences on fundamental issues between post-1935 interpretations of quantum mechanics by Schrödinger, Everett, and others, see Bitbol 1996 and Perovic 2003.

CHAPTER 11

1. Some of the ideas and arguments presented here have been previously developed in Perovic 2008, in response to Muller's (1999, 1997a, 1997b) account of Schrödinger's proof of equivalence of wave and matrix mechanics. Here I put them in the more general context of the main ideas developed in the book.

2. See a very insightful analysis by Dieks (2017) on the notion of "symbolic forms" in Bohr's work.

3. Perhaps both formalism and complementarity were much too provisional at the time of their early development to be understood by excessive conceptual dissection in which clarity was achieved by the anachronistic imposition of contemporary terms, though that sort of analysis may have a certain value.

4. For a more detailed analysis, see Perovic 2008.

5. Muller's notion of mathematical equivalence is much stronger than proved equivalence. Though the latter employs mathematical techniques, it has no explicit goal of arriving at a conclusion about the logical structures of the theories. Muller writes, "The essence of a physical theory lies in the mathematical structures it employs; to describe physical systems, the equivalence proof . . . can legitimately be construed as an attempt to demonstrate the isomorphism between the mathematical structures of Matrix Mechanics and Wave Mechanics" (Muller 1997a, 38). While Muller never explicitly claims that isomorphism of matrix mechanics and wave mechanics was Schrödinger's main goal, his reconstruction of the proof, in which "Matrix Mechanics and Wave Mechanics such as they were around March 1926 are thus tailored in structural terms" (ibid., 38), indicates that the proof's goal could not be much different.

CHAPTER 12

1. E. Scheibe (1973) similarly interprets complementarity as an account focusing on complementarity of *phenomena*—the pieces of information or forms of experience that supposedly avoid the ontological contradictions and dilem-

mas of classical approaches, thereby allowing us to use the classical properties selectively in interpreting experiments.

2. Bohr also points out Heisenberg's interesting suggestion that this is applicable to macrostates as well as "macroscopic phenomena . . . in a certain sense . . . created by repeated observations" (Bohr 1928, 584).

3. See more on Heisenberg's interpretation of Bohr's complementarity in Camilleri 2009, 112–16.

4. Joas and Lehner (2009) take up the case of the application of quantum mechanics to molecular phenomena.

5. I developed a detailed historical and philosophical account of quantum tunneling with respect to Bohr's method in Perović 2017. Here I briefly summarize the main results of that analysis, and explain how it relates to the key argument in part 3 of this book.

CHAPTER 13

1. The two kinds of analysis, historically and formally oriented, of this particular case in the history of quantum physics are discussed by Vickers (2014).

2. For a recent informative discussion of the notion of collapse in the context of Bohr's work, see Zinkernagel 2016.

CHAPTER 14

1. From Ehrenfest's letter to Einstein dated 16 September 1925. See the content of the letter at https://einsteinpapers.press.princeton.edu/vol15-doc/234.

2. In another passage on the subject, he says: "The question was whether, as to the occurrence of individual effects, we should adopt a terminology proposed by Dirac, that we were concerned with a choice on the part of 'nature' or, as suggested by Heisenberg, we should say that we have to do with a choice on the part of the 'observer' constructing the measuring instruments and reading their recording. Any such terminology would, however, appear dubious since, on the one hand, it is hardly reasonable to endow nature with volition in the ordinary sense, while, on the other hand, it is certainly not possible for the observer to influence the events which may appear under the conditions he has arranged. To my mind, there is no other alternative than to admit that, in this field of experience, we are dealing with individual phenomena and that our possibilities of handling the measuring instruments allow us only to make a choice between the different complementary types of phenomena we want to study" (Bohr 1949, 223).

CHAPTER 15

1. After all, Newton told us that he had never used hypotheses to develop his grand theory of motion. He claimed that he had instead induced it straightforwardly from the laws Kepler had formulated, and from additional empirical

evidence. He had very strong views on his method, but those views could have been incorrect. He developed the theoretical notion of force such that the mass of the planets and the sun was accounted for differently than in Kepler's laws (the difference between the mass of the sun and that of the planets matters in Newton's laws, among other things). So it is possible that he was incorrect about the nature of his own method. See, e.g., Feyerabend 1993.

BIBLIOGRAPHY

Aaserud, F., and J. L. Helibron. 2013. *Love, Literature and the Quantum Atom.* Oxford: Oxford University Press.

Achinstein, P. 1974. History and Philosophy of Science: A Reply to Cohen. In *The Structure of Scientific Theories*, edited by F. Suppe. Urbana: University of Illinois Press, 350–60.

———. 1993. How to Defend a Theory without Testing It: Niels Bohr and the "Logic of Pursuit." *Midwest Studies in Philosophy* 18, no. 1: 90–120.

Aspect, A., P. Grangier, and G. Roger. 1982. Experimental Realization of Einstein-Podolsky-Rosen-Bohm Gedankenexperiment: A New Violation of Bell's Inequalities. *Physical Review Letters* 49, no. 2: 91.

Bacciagaluppi, G. 2012. The Role of Decoherence in Quantum Mechanics. In *The Stanford Encyclopedia of Philosophy*, edited by E. N. Zalta. Accessed at http://plato.stanford.edu/archives/win2012/entries/qmdecoherence.

Bacon, F. 1874. *The Works of Francis Bacon*, 14 vols., edited by J. Spedding, R. L. Ellis, and D. D. Heath. London: Longman.

———. 2000. *Francis Bacon: The New Organon.* Cambridge, New York, and Melbourne: Cambridge University Press.

Balmer, J. J. 1885. Notiz über die Spectrallinien des Wasserstoffs. *Annalen der Physik* 261, no. 5: 80–87.

Bächtold, M. 2017. On Bohr's Epistemological Contribution to the Quantum-Classical Cut Problem. *Niels Bohr and the Philosophy of Physics: Twenty-First-Century Perspectives*, edited by J. Faye and H. J. Folse, 235–52. London and New York: Bloomsbury Academic.

Becker, A. 2018. *What Is Real? The Unfinished Quest for the Meaning of Quantum Physics.* London: Basic Books.

Bell, J. S. 1964. On the Einstein-Podolsky-Rosen Paradox. *Physics* 1, no. 3: 195–290.

———. 2001. Six Possible Worlds of Quantum Mechanics. In *John S. Bell on the Foundations of Quantum Mechanics*, edited by M. Bell, K. Gottfried, and M. Veltman. River Edge, NJ: World Scientific.

Beller, M. 1992. The Birth of Bohr's Complementarity: The Context and the Dialogues. *Studies in History and Philosophy of Science Part A* 23, no. 1: 147–80.

———. 1997. Against the Stream: Schrödinger's Interpretation of Quantum Mechanics. *Studies in History and Philosophy of Modern Physics* 28, no. 3: 424.

———. 1999. *Quantum Dialogue: The Making of a Revolution.* Chicago: University of Chicago Press.

Bitbol, M. 1995. Introduction to E. Schrödinger, *The Interpretation of Quantum Mechanics*, edited by M. Bitbol. Woodbridge, CT: Ox Bow Press.

———. 1996. Schrödinger's Philosophy of Quantum Mechanics. Dodrecht, Netherlands: Kluwer.

———. 2017. On Bohr's Transcendental Research Program. In *Niels Bohr and the Philosophy of Physics: Twenty-First-Century Perspectives*, edited by J. Faye and H. J. Folse, 47–66, London and New York: Bloomsbury.

Bloor, D. 1991. *Knowledge and Social Imagery*. Chicago: University of Chicago Press.

Bohm, D. 1957. *Causality and Chance in Modern Physics*. London: Routledge and Kegan Paul.

Bohr, N. 1909. Determination of the Surface-Tension of Water by the Method of Jet Vibration. *Philosophical Transactions of the Royal Society of London A* 209:281–317.

———. 1913a. On the Constitution of Atoms and Molecules. *London, Edinburgh, and Dublin Philosophical Magazine and Journal of Science* 26, no. 151: 1–25.

———. 1913b. The Spectra of Helium and Hydrogen. *Nature* 92:231–32.

———. 1913c. On the Constitution of Atoms and Molecules. *London, Edinburgh, and Dublin Philosophical Magazine and Journal of Science* 26, no. 153: 476–502.

———. 1913d. On the Constitution of Atoms and Molecules. *London, Edinburgh, and Dublin Philosophical Magazine and Journal of Science* 26, no. 155: 857–75.

———. 1915. On the Quantum Theory of Radiation and the Structure of the Atom. *London, Edinburgh, and Dublin Philosophical Magazine and Journal of Science* 30, no. 177: 394–415.

———. 1922a. *The Theory of Spectra and Atomic Constitution: Three Essays*. Cambridge: Cambridge University Press.

———. 1922b. Der Bau der Atome und die physikalischen und chemischen Eigenschaften der Elemente. *Zeitschrift für Physik* 9:1–67.

———. 1922c. Letter printed in *Københavens Universitetet Aarbog*, 1915–1920, part IV, 283–87. English translation in A. Pais, A. 1989. "Physics in the Making in Bohr's Copenhagen." In *Physics in the Making*, edited by A. Sarlemijn and M. J. Sparnaaj. Amsterdam, Oxford, New York, and Tokyo: North Holland.

———. 1924. *On the Application of the Quantum Theory to Atomic Structure: Part I. The Fundamental Postulates*. Cambridge: Cambridge University Press.

———. 1925. Atomic Theory and Mechanics. *Nature* 116, no. 2927: 845–52.

———. 1928. The Quantum Postulate and the Recent Development of Atomic Theory. *Nature* (supplement), April 14: 580–90.

———. 1931. Maxwell and Modern Theoretical Physics. *Nature* 128:691–92.

———. 1934. *Atomic Theory and the Description of Nature*. Cambridge: Cambridge University Press.

———. 1935. Can Quantum-Mechanical Description of Physical Reality Be Considered Complete? *Physical Review* 48, no. 8: 696–702.

———. 1939. Natural Philosophy and Human Cultures. *Nature* 143:268–72.

———. 1948. On the Notions of Causality and Complementarity. *Dialectica* 2:312–19.

———. 1949. Discussion with Einstein on Epistemological Problems in Atomic Physics. In *Albert Einstein, Philosopher-Scientist*, edited by P. A. Shlipp, 199–242. Evanston, IL: Library of Living Philosophers.

———. 1958a. Quantum Physics and Philosophy: Causality and Complementarity. In *Philosophy at Mid-Century: A Survey*, edited by R. Klubansky, 308–14. Florence: La Nuova Italia Editrice.

———. 1958b. On Atoms and Human Knowledge. *Daedalus* 87, no. 2: 164–75.

———. 1961. The Rutherford Memorial Lecture 1958 Reminiscences of the Founder of Nuclear Science and of Some Developments Based on his Work. *Proceedings of the Physical Society* 78, no. 6: 1083–1115.

———. 1962. Interview of N. Bohr by T. S. Kuhn, L. Rosenfeld, E. Riidinger, and A. Petersen, October 31 and November 7, 1962.

———. 1972–2008. *Niels Bohr: Collected Works*, vols. 1–13. Amsterdam: North Holland.

Bohr, N., H. A. Kramers, and J. C. Slater. 1924a. Über die Quantentheorie der Strahlung. *Zeitschrift für Physik* 24:69–87.

———. 1924b. The Quantum Theory of Radiation. *Philosophical Magazine* 47:785–802.

Bohr, N., and L. Rosenfeld. 1933. Zur Frage der Masbarkeit der elektromgnetischen Feldgressen. In *Selected Papers of Leon Rosenfeld*, translated by R. Cohen and J. Stachel, 357–400. Dordrecht, Netherlands: Springer, 1979.

Bohr, N., and J. A. Wheeler. 1939. The Mechanism of Nuclear Fission. *Physical Review* 56, no. 5: 426.

Bokulich, P., and A. Bokulich. 2005. Niels Bohr's Generalization of Classical Mechanics. *Foundations of Physics* 35, no. 3: 347–71.

Born, M. 1953. The Interpretation of Quantum Mechanics. *British Journal for the Philosophy of Science* 4:95–106.

———. 1961. Bemerkungen zur statistischen Deutung der Quantenmechanik. In *Werner Heisenberg und die Physik unserer Zeit*, 103. Braunschweig, Germany: Vieweg.

Born, M., W. Heisenberg, and P. Jordan. 1926. Zur Quantenmechanik. II. *Zeitschrift für Physik* 35, nos. 8–9: 557–615.

Bothe, W., and H. Geiger. 1925. Über das Wesen des Comptoneffekts: Ein experimenteller Beitrag zur Theorie der Strahlung. *Zeitschrift für Physik* 32, no. 1: 639–63.

Boyd, N. M. 2018. Evidence Enriched. *Philosophy of Science* 85, no. 3: 403–21.

Bub, J. 1974. *The Interpretation of Quantum Mechanics.* Dordrecht, Netherlands, and Boston: D. Reidl.

———. 1977. Reply to Professor Causey. In *The Structure of Scientific Theories*, edited by F. Suppe, 402–8. Urbana: University of Illinois Press.

Buchwald, J. Z. 1995. Why Hertz Was Right about Cathode Rays. In *Scientific*

Practice: Theories and Stories of Doing Physics, 151–70. Chicago: University of Chicago Press.

Buchwald, J. Z., and A. Warwick, eds. 2001. *Histories of the Electron: The Birth of Microphysics*. Cambridge, MA: MIT Press.

Bueno, O., and P. Vickers. 2014. Is Science Inconsistent? *Synthese* 191, no. 13: 2887–89.

Burian, R. M. 2001. The Dilemma of Case Studies Resolved: The Virtues of Using Case Studies in the History and Philosophy of Science. *Perspectives on Science* 9, no. 4: 383–404.

———. 2002. Comments on the Precarious Relationship between History and Philosophy of Science. *Perspectives on Science* 10, no. 4: 398–407.

Büttiker, M. and S. Washburn. 2003. Optics: Ado about Nothing Much? *Nature* 422, no. 6929: 271–72.

Camilleri, K. 2009. *Heisenberg and the Interpretation of Quantum Mechanics*. Cambridge: Cambridge University Press.

———. 2017. Why Do We Find Bohr Obscure? Reading Bohr as a Philosopher of Experiment. In Fay J. and Folse H.J., eds., *Niels Bohr and the Philosophy of Physics: the 21ˢᵗ Century Perspective.*, edited by J. Fay and H. J. Folse. London and New York: Bloomsbury.

Camilleri, K., and M. Schlosshauer. 2015. Niels Bohr as Philosopher of Experiment: Does Decoherence Theory Challenge Bohr's Doctrine of Classical Concepts? *Studies in History and Philosophy of Science Part B: Studies in History and Philosophy of Modern Physics* 49:73–83.

Carazza, B., and N. Robotti. 2002. Explaining Atomic Spectra within Classical Physics: 1897–1913. *Annals of Science* 59, no. 3: 299–320.

Casimir, H. B. G. 1968. Recollection from the Years 1929–1931. In *Niels Bohr: His Life and Work as Seen by His Friends and Colleagues*, edited by S. Rozental, 109–13. Amsterdam: North-Holland, and New York: Wiley.

Cassidy, D. C. 1979. Heisenberg's First Core Model of the Atom: The Formation of a Professional Style. *Historical Studies in the Physical Sciences* 10:187–224.

Cassirer, E. 1956. *Determinism and Indeterminism in Modern Physics Historical and Systematic Studies of the Problem of Causality*. New Haven: Yale University Press.

Chalmers, A. 2011. Drawing Philosophical Lessons from Perrin's Experiments on Brownian Motion: A Response to Van Fraassen. *British Journal for the Philosophy of Science* 62, no. 4: 711–32.

Chang, H. 2004. *Inventing Temperature: Measurement and Scientific Progress*. Oxford, UK: Oxford University Press.

Chen, R. L. 2007. The Structure of Experimentation and the Replication Degree: Reconsidering the Replication of Hertz's Cathode Ray Experiment. *Rodopi Philosophical Studies* 7:129–49.

Chevalley, C. 1994. "Niels Bohr's Words and the Atlantis of Kantianism". In J. Faye and H. Folse, eds., *Niels Bohr and Contemporary Philosophy* (series *Boston Studies in History and Philosophy of Science*, vol. 153), 33–55. Dordrecht, Netherlands: Kluwer.

Chiao, R. Y. 1998. Tunneling Times and Superluminality: A Tutorial. Accessed at https://arxiv.org/pdf/quant-ph/9811019.pdf.

Clark, K. J. 2014. Judaism and Evolution. In *Religion and the Sciences of Origins*, 207–22. New York: Palgrave Macmillan.

Compton, A. H. 1922a. Secondary Radiations Produced by X-rays, and Some of Their Applications to Physical Problems. *Bulletin of the National Research Council*, 4, no. 20.

———. 1922b. Radiation a Form of Matter. *Science* 56:716–17.

———. 1923a. Wave-Length Measurements of Scattered X-rays. *Physical Review* 21:715.

———. 1923b. A Quantum Theory of the Scattering of X-rays by Light Elements. *Physical Review* 21:484.

———. 1923c. The Total Reflection of X-rays. *Philosophical Magazine* 45:1121–31.

———. 1923d. Recoil of Electrons from Scattered X-rays. *Nature* 112:435.

Compton, A. H., and A. W. Simon. 1925. Directed Quanta of Scattered X-rays. *Physical Review* 26:289–99.

Condon, E. U., and P. M. Morse. 1931. Quantum Mechanics of Collision Processes I: Scattering of Particles in a Definite Force Field. *Reviews of Modern Physics* 3, no. 1: 43.

Conway, A. W. 1907. On Series in Spectra. *Scientific Proceedings of the Royal Dublin Society* 9:51.

Cuffaro, M. 2010. The Kantian Framework of Complementarity. *Studies in History and Philosophy of Science Part B: Studies In History and Philosophy of Modern Physics* 41, no. 4: 309–17.

Cushing, J. T. 1994. *Quantum Mechanics: Historical Contingency and the Copenhagen Hegemony*. Chicago: University of Chicago Press.

Darrigol, O. 1992. *From c-Numbers to q-Numbers*. Berkeley: University of California Press.

Darwin, C. G. 1912. A Theory of the Absorption and Scattering of the α Rays. *London, Edinburgh, and Dublin Philosophical Magazine and Journal of Science* 23, no. 138: 901–20.

D'Abro, A. 1951. *The Rise of the New Physics*. New York: Dover Publications.

De Chadarevian, S., and N. Hopwood. 2004. *Models: The Third Dimension of Science*. Stanford, CA: Stanford University Press.

De Gregorio, A. 2014. Bohr's Way to Defining Complementarity. *Studies in History and Philosophy of Science Part B: Studies in History and Philosophy of Modern Physics* 45:72–82.

Dickson, M. 2004. Quantum Reference Frames in the Context of EPR. *Philosophy of Science* 71, no. 5: 655–68.

Dieks, D. 2017. Niels Bohr and the Mathematical Formalism. In *Niels Bohr and the Philosophy of Physics: Twenty-First-Century Perspectives*, edited by J. Faye and H. J. Folse, 303–34. London and New York: Bloomsbury.

Dirac, P. 1927. The Quantum Theory of the Emission and Ansorption of Radiation. *Proceedings of the Royal Society* 114 A: 243–65.

———. 1930. *The Principles of Quantum Mechanics*. Oxford: Clarendon Press.

Dorato, M. 2017. Classical-Quantum Interaction. In *Niels Bohr and the Philosophy of Physics: Twenty-First-Century Perspectives*, edited by J. Faye and H. J. Folse, 133–55. London and New York: Bloomsbury.

Duhem, P. 1954. *The Aim and Structure of Physical Theory*. Princeton, NJ: Princeton University Press.

Duncan, A., and M. Janssen. 2009. From Canonical Transformations to Transformation Theory, 1926–1927: The Road to Jordan's Neue Begründung. *Studies in History and Philosophy of Science Part B: Studies In History and Philosophy of Modern Physics* 40, no. 4: 352–62.

Earman, J. 1992. *Bayes or Bust? A Critical Examination of Bayesian Confirmation Theory*. Cambridge, MA: MIT Press.

Eckart, C. 1926. Operator Calculus and the Solution of the Equations of Motion of Quantum Dynamics. *Physical Review* 28:711–26.

Einstein, A. 1917. Zur Quantentheorie der Strahlung. *Physikalische Zeitschrift* 18:121–28.

———. 1949. Autobiographical Notes. In *Albert Einstein: Philosopher-Scientist*, edited by Paul A. Schipp, 1–95. New York: Library of Living Philosophers.

Einstein, A., B. Podolsky, and N. Rosen. 1935. Can Quantum-Mechanical Description of Physical Reality Be Considered Complete? *Physical Review* 47, no. 10: 777–80.

Elgin, C. Z. 1996. *Considered Judgment*. Princeton, NJ: Princeton University. Press.

Esfeld, M. 2019. Against the Disappearance of Spacetime in Quantum Gravity. *Synthese*, Accessed at https://doi.org/10.1007/s11229-019-02168-y.

Favrholdt, D. 1992. Niels Bohr's Philosophical Background. In *Historisk-filosofiske meddelelser*, vol. 63. Copenhagen: Munksgaard.

Faye, J. 1991. *Niels Bohr: His Heritage and Legacy; An Anti-Realist View of Quantum Mechanics*. Dordrecht, Netherlands: Kluwer.

———. 2017. Complementarity and Human Nature. *Niels Bohr and the Philosophy of Physics: Twenty-First-Century Perspectives*, edited by J. Faye and H. J. Folse, 115–32. London and New York: Bloomsbury.

Feyerabend, P. K. 1969. On a Recent Critique of Complementarity: Part II. *Philosophy of Science* 36, no. 1: 82–105.

———. 1970. *Against Method: Outline of an Anarchistic Theory of Knowledge*. Minneapolis: University of Minnesota Press.

———. 1993. *Against Method*. London: Verso.

Fowler, A. 1914. Bakerian Lecture: Series Lines in Spark Spectra. *Proc. R. Soc. Lond. A* 90, no. 620: 426–30.

Fowler, R. H., and L. Nordheim 1928. Electron Emission in Intense Electric Fields. *Proceedings of the Royal Society of London A: Mathematical, Physical and Engineering Sciences* 119, no. 781: 173–81.

Franck, J., and G. Hertz. 1913. Über Zusammenstöße zwischen Gasmolekülen und langsamen Elektronen. *Verhandlungen der Deutschen Physikalischen Gesellschaft* 15:373–90.

———. 1919. Die Bestätigung der Bohrschen Atomtheorie im optischen Spektrum durch Untersuchungen der unelastischen Zusammenstöße langsamer Elektronen mit Gasmolekülen. *Physikalische Zeitschrift* 20:132–43.

Franklin, A. 1984. Are Paradigms Incommensurable? *British Journal for the Philosophy of Science* 35, no. 1: 57–60.

———. 1989. *Neglect of Experiment*. Cambridge, New York, and Melbourne: Cambridge University Press.

Franklin, A., and C. Howson. 1984. Why Do Scientists Prefer to Vary Their Experiments? *Studies in History and Philosophy of Science Part A* 15, no. 1: 51–62.

Franklin, A., and S. Perovic. 1998. Experiments in Physics. In *Stanford Encyclopedia of Philosophy*. Accessed at https://plato.stanford.edu/entries/physics-experiment/.

Frisch, O. R. 1968. The Interest Is Focusing on the Atomic Nucleus. In *Niels Bohr: His Life and Work as Seen by His Friends and Colleagues*, edited by S. Rozental, 137–48. Amsterdam: North-Holland, and New York: Wiley.

Galison, P. 1987. *How Experiments End*. Chicago: University of Chicago Press.

———. 1997. *Image and Logic: A Material Culture of Microphysics*. Chicago: University of Chicago Press.

Gamow, G. 1928. Zur Quantentheorie des Atomkernes. *Zeitschrift für Physik* 51, nos. (3-4), 204–12.

———. 1966. *Thirty Years that Shook Physics: The Story of Quantum Theory*. Chelmsford, MA: Courier Corporation.

Gibbins, P. 1987. *Particles and Paradoxes*. Cambridge: Cambridge University Press.

Giere, R. N. 1973. History and Philosophy of Science: Intimate Relationship or Marriage of Convenience? *Minnesota Studies in the Philosophy of Science* 24, no. 3: 282–97.

———. 2010. *Scientific Perspectivism*. University of Chicago Press.

———. 2011. History and Philosophy of Science: Thirty-Five Years Later. In *Integrating History and Philosophy of Science*, 59–65. Dordrecht, Netherlands: Springer.

Gooding, D. 1990. Mapping Experiment as a Learning Process: How the First Electromagnetic Motor Was Invented. *Science, Technology & Human Values* 15, no. 2: 165–201.

Gurney, R. W., and E. U. Condon. 1928. Wave Mechanics and Radioactive Disintegration. *Nature* 122, no. 3073: 439.

———. 1929. Quantum Mechanics and Radioactive Disintegration. *Physical Review* 33, no. 2: 27.

Hacking, I. 1974. *The Emergence of Probability*. Cambridge, New York, and Melbourne: Cambridge University Press.

Hanle, P. A. 1977. The Coming of Age of Erwin Schrödinger: His Quantum Statistics of Ideal Gases. *Archive for History of Exact Sciences* 17, no. 2: 165–92.

Hanson, N. R. 1963. *The Concept of the Positron*. Cambridge: Cambridge University Press.

Harré, R. 1981. *Great Scientific Experiments*. Oxford: Oxford University Press.

Hattiangadi, J. 2006. On the True Method of Induction or Investigative Induction. Accessed at http://philsci-archive.pitt.edu/3109/.

Heilbron, J. L. 2013. The Mind That Created the Bohr Atom. In N. Bohr, *Poincaré Séminaire*, vol. 17, edited by O. Darrigol, B. Duplantier, J. M. Raimond, and V. Rivasseau, 19–58. Basel, Switzerland: Birkhäuser.

Heilbron, J. L., and T. S. Kuhn. 1969. The Genesis of the Bohr Atom. *Historical Studies in Physical Sciences* 1:211–90.

Heisenberg, W. 1925. Über quantentheoretische Umdeutung kinematischer und mechanischer Beziehungen. *Zeitschrift für Physik* 33, no. 1: 879–93.

———. 1955. The Development of the Interpretation of the Quantum Theory. In *Niels Bohr and the Development of Physics*, edited by W. Pauli, 24. New York: McGraw Hill.

———. 1968. Quantum Theory and Its Interpretation. In *Niels Bohr: His Life and Work as Seen by His Friends and Colleagues,* edited by S. Rozental, 94–108. Amsterdam: North-Holland, and New York: Wiley.

———. 1983. The Physical Content of Quantum Kinematics and Mechanics. In *Quantum Theory and Measurement*, edited by J. A. Wheeler and W. H. Zurek, 62–84. Princeton, NJ: Princeton University Press.

Heisenberg, W., and B. Bond. 1959. *Physics and Philosophy: The Revolution in Modern Science*. St. Leonards, Australia: Allen & Unwin.

Heitler, W., and F. London. 1927. Wechselwirkung neutraler Atome und homöopolare Bindung nach der Quantenmechanik. *Zeitschrift für Physik* 44, nos. 6-7: 455–72.

Held, C. 1994. The Meaning of Complementarity. *Studies in History and Philosophy of Science Part A* 25, no. 6: 871–93.

Henderson, L. 2018. The Problem of Induction. *Stanford Encyclopedia of Philosophy*. Accessed at https://plato.stanford.edu/entries/induction-problem/.

Hertz, H. 1883. Versuche über die Glimmentladung, *Annalen der Physik und Chemie* 19:782–816. Translated as Experiments on the Cathode Discharge, in Heinrich Hertz, *Miscellaneous Papers*, 224–54. London: Macmillan and Company.

Hettema, H. 1995. Bohr's Theory of the Atom 1913–1923: A Case Study in the Progress of Scientific Research Programmes. *Studies in History and Philosophy of Modern Physics* 26:307–23.

Hilgevoord, J., and J. Uffink. 2016. The Uncertainty Principle. In *The Stanford Encyclopedia of Philosophy* (winter 2016 edition), edited by Edward N. Zalta. Accessed at https://plato.stanford.edu/archives/win2016/entries/qt-uncertainty/.

Holst, H., R. B. Lindsay, R. T. E. Lindsay, and H. A. Kramers. 1923. The Atom and the Bohr Theory of Its Structure: An Elementary Presentation. London: Gyldendal.

Hooker, C. A. 1972. The Nature of Quantum Mechanical Reality. In *Paradigms and Paradoxes*, edited by R. G. Colodny, 67–305. Pittsburgh: University of Pittsburgh Press.

Hossenfelder, S. 2018. *Lost in Math: How Beauty Leads Physics Astray*. London: Basic Books.

Howard, D. 1979. Complementarity and Ontology: Niels Bohr and the Problem of Scientific Realism in Quantum Physics, Ph.D. dissertation, Boston University.

———. 1994. What Makes a Classical Concept Classical? In *Niels Bohr and Contemporary Philosophy*, 201–29. Dordrecht, Netherlands: Springer.

———. 2004. Who Invented the "Copenhagen Interpretation"? A Study in Mythology. *Philosophy of Science* 71, no. 5: 669–82.

———. 2007. Revisiting the Einstein-Bohr Dialogue. *Iyyun: The Jerusalem Philosophical Quarterly /עיון: רבעון פילוסופי*. Accessed at https://www3.nd.edu/~dhoward1/Revisiting%20the%20Einstein-Bohr%20Dialogue.pdf.

Huggett, N., and C. Wüthrich. 2013. Emergent Spacetime and Empirical (in) Coherence. *Studies in History and Philosophy of Science Part B: Studies in History and Philosophy of Modern Physics* 44, no. 3: 276–85.

Hughes, J. 1998. "Modernists with a Vengeance": Changing Cultures of Theory in Nuclear Science, 1920–1930. *Studies in History and Philosophy of Modern Physics* 29:339–67.

Hund, F. 1927a. Zur Deutung der Molekelspektren I. *Zeitschrift für Physik* 40, no. 10: 742–64.

———. 1927b. Zur Deutung der Molekelspektren II. *Zeitschrift für Physik* 42, nos. 2-3: 93–120.

———. 1927c. Zur Deutung der Molekelspektren III. *Zeitschrift für Physik* 43, nos. 11-12: 805–26.

Hutten, E. H. 1956. On Explanation in Psychology and in Physics. *British Journal for the Philosophy of Science* 7, no. 25:73–85.

James, J., and C. Joas. 2015. Subsequent and Subsidiary? Rethinking the Role of Applications in Establishing Quantum Mechanics. *Historical Studies in the Natural Sciences* 45, no. 5: 641–702.

Jammer, M. 1966. *The Conceptual Development of Quantum Mechanics*. New York: McGraw-Hill.

———. 1974. *Philosophy of Quantum Mechanics: The Interpretations of Quantum Mechanics in Historical Perspective*. New York: Wiley.

———. 1989. *Conceptual Development of Quantum Mechanics*. New York: American Institute of Physics.

Jeans, J. H. 1901. The Mechanism of Radiation. *Philosophical Magazine* 2:421–55.

Johansson, L. G. 2007. *Interpreting Quantum Mechanics: A Realistic View in Schrodinger's Vein*. London and New York: Routledge.

Joas, C., and S. Katzir. 2011. Analogy, Extension, and Novelty: Young Schrödinger on Electric Phenomena in Solids. *Studies in History and Philoso-*

phy of Science Part B: Studies in History and Philosophy of Modern Physics 42, no. 1: 43–53.

Joas, C., and C. Lehner. 2009. The Classical Roots of Wave Mechanics: Schrödinger's Transformations of the Optical-Mechanical Analogy. *Studies in History and Philosophy of Science Part B: Studies in History and Philosophy of Modern Physics* 40, no. 4: 338–51.

Joos, E. 2006. The Emergence of Classicality from Quantum Theory. In *The ReEmergence of Emergence: The Emergentist Hypothesis from Science to Religion*, edited by P. Clayton and P. Davies. Oxford, UK: Oxford University Press.

Joos, E., H. D. Zeh, C. Kiefer, D. Giulini, J. Kupsch, and I.-O. Stamatescu. 2003. *Decoherence and the Appearance of a Classical World in Quantum Theory*, 2nd edition. New York: Springer.

Jordan, P. 1944. *Physics of the Twentieth Century*. New York: Philosophical Library.

Kaiser, D. 1992. More Roots of Complementarity: Kantian Aspects and Influences. *Studies in History and Philosophy of Science* 23:213–39.

Kangro, H. 1976. *Early History of Planck's Radiation Law*. London: Taylor & Francis.

Karaca, K. Forthcoming. Two Senses of Experimental Robustness: Result Robustness and Procedure Robustness. *British Journal for the Philosophy of Science*.

Katsumori, M. 2011. *Niels Bohr's Complementarity: Its Structure, History, and Intersections with Hermeneutics and Deconstruction*, Boston Studies in History and Philosophy of Science, vol. 286. Dordrecht, Heidelberg, London, and New York: Springer Science & Business Media.

Kauark-Lite, P. 2017. Transcendental versus Quantitative Meanings of Bohr's Complementarity. In *Niels Bohr and the Philosophy of Physics: Twenty-First-Century Perspectives*, edited by J. Faye and H. J. Folse, 67–90, London and New York: Bloomsbury.

Kidd, R., J. Ardini, and A. Anton. 1985. Compton Effect as a Double Doppler Shift. *American Journal of Physics* 53:641–44.

Klein, M. J. 1964. Einstein and the Wave-Particle Dduality. *The Natural Philosopher* 3:1–49.

———. 1968. Glimpses of Niels Bohr as Scientist and Thinker. In *Niels Bohr: His Life and Work as Seen by His Friends and Colleagues,* edited by S. Rozental, 74–93. Amsterdam: North-Holland, and New York: Wiley.

Kossel, W. 1914. Bemerkung zur absorption homogener Röntgenstrahlen. *Verhandlungen der Deutschen Physikalischen Gesellschaft* 16:898–909.

Kragh, H. 1979. Niels Bohr's Second Atomic Theory. *Historical Studies in the Physical Sciences* 10:123–86.

———. 1982. Erwin Schrödinger and the Wave Equation: The Crucial Phase. *Centaurus* 26, no. 2: 154–97.

———. 1985. The Fine Structure of Hydrogen and the Gross Structure of the Physics Community, 1916–26. *Historical Studies in the Physical Sciences* 15, no. 2: 67–125.

———. 2012. *Niels Bohr and the Quantum Atom: The Bohr Model of Atomic Structure 1913–1925*. Oxford, UK: Oxford University Press.

———. 2013. The Many Faces of the Bohr Atom. Accessed at https://arxiv.org/abs/1309.4200.

Kramers, H. A. 1923. Das Korrespondenzprinzip und der Schalenbau des Atoms. *Naturwissenschaften* 11, no. 27: 550–59.

———. 1935. Atom-og Kvanteteoriens Udvikling i Arene 1913–1925. *Fysisk tidsskrift* 33:82–96.

Kramers, H. A., H. Holst, R. B. Lindsay, and R. T. Lindsay. 1923. *The Atom and the Bohr Theory of Its Structure*, 133. London: Gyldendal.

Kuhn, T. S. 1970. *The Structure of Scientific Revolutions.* Chicago: University of Chicago Press.

———. 1963. Interview with Werner Heisenberg 02/11/1963. Niels Bohr Library and Archives, American Institute of Physics, College Park, MD. www.aip.org/history-programs/niels-bohr-library/oral-histories/4661-8.

———. 1987. *Black-Body Theory and the Quantum Discontinuity, 1894–1912*. Chicago: University of Chicago Press.

Lakatos, I. 1970. Falsification and the Methodology of Scientific Research Programmes. In *Criticism and the Growth of Knowledge* vol. 4, edited by A. Musgrave and I. Lakatos, 91–196, Cambridge: Cambridge University Press.

Landé, A. 1962. Interview by T. S. Kuhn and J. L. Heilbron, 5 March 1962.

Landsman, N. P. 2006. When Champions Meet: Rethinking the Bohr-Einstein Debate. *Studies in History and Philosophy of Modern Physics* 37:212–42.

Langevin, P., and M. de Broglie. 1912. *La théorie du rayonnement et les quanta: Rapports et discussions de la réunion tenue à Bruxelles, du 30 Octobre au 3 Novembre 1911, sous les auspices de E. Solvay (No. 7).* Paris: Gauthier-Villars.

Larmor, J. 1900. *Aether and Matter*. Cambridge: Cambridge University Press.

Laudan, L. 1977. *Progress and Its Problems*. Berkeley: University of California Press.

Laymon, R. 1994. Demonstrative Induction, Old and New Evidence and the Accuracy of the Electrostatic Inverse Square Law. *Synthese* 99, no. 1: 23–58.

Lenard, P. 1903. Über die Absorption von Kathodenstrahlen verschiedener Geschwindigkeit. *Annalen der Physik* 317, no. 12: 714–44.

Lloyd, E. A. 2012. The Role of "Complex" Empiricism in the Debates about Satellite Data and Climate Models. *Studies in History and Philosophy of Science Part A* 43, no. 2: 390–401.

MacKinnon, E. 1985. Bohr on the Foundations of Quantum Theory. In *Niels Bohr: A Centenary Volume*, edited by P. J. Kennedy, 101–20. Cambridge, MA: Harvard University Press..

Mattingly, J. 2001. The Replication of Hertz's Cathode Ray Experiments. *Studies in History and Philosophy of Science Part B: Studies in History and Philosophy of Modern Physics* 32, no. 1: 53–75.

Maudlin, T. 2018. Does Anyone Understand Quantum Physics and Its Implications on Reality? *Quora*. Accessed at https://www.quora.com/Does-anyone

-understand-quantum-physics-and-its-implications-on-reality/answer/Tim
-Maudlin-1?share=a5141392&srid=5CTnL.

———. 2019. *Philosophy of Physics: Quantum Theory.* Princeton, NJ, and Oxford, UK: Princeton University Press.

McLaren, S. B. 1913. IV. The Theory of Radiation. *The London, Edinburgh, and Dublin Philosophical Magazine and Journal of Science* 25, no. 145: 43–56.

McMullin, E. 1954. *The Principle of Uncertainty.* (Unpublished PhD dissertation, Catholic University of Louvain.

———. 1970. The History and Philosophy of Science: A Taxonomy. In *Minnesota Studies in the Philosophy of Science,* vol. 5, edited by R. Stuewer, 12–67. Minneapolis: University of Minnesota Press.

———. 1990. Conceptions of Science in the Scientific Revolution. In *Reappraisals of the Scientific Revolution,* edited by D. Lindberg and R. S. Westman, 27–92. Cambridge: Cambridge University Press.

Mehra, J., and H. Rechenberg. 1982. The Historical Development of Quantum Theory. New York: Pergamon Press.

Merzbacher, E. 2002. The Early History of Quantum Tunneling. *Physics Today* 55, no. 8: 44–50.

Millikan, R. A. 1916. A Direct Photoelectric Determination of Planck's *h*. *Physical Review* 7:355–88.

Morgan, M. S., and M. Morrison. 1999. *Models as Mediators.* Cambridge: Cambridge University Press.

Moseley, H. G. J. 1914. Atomic Models and X-ray Spectra. *Nature* 92, no. 2307: 554.

Mott, N. F. 1929. The Wave-Mechanics of A-ray Tracks. *Proceedings of Royal Society* A126:79–84.

Muller, F. A. 1997a. The Equivalence Myth of Quantum Mechanics, Part I. *Studies in History and Philosophy of Science Part B: Studies in History and Philosophy of Modern Physics* 28, no. 1: 35–61.

———. 1997b. The Equivalence Myth of Quantum Mechanics, Part II. *Studies in History and Philosophy of Science Part B: Studies in History and Philosophy of Modern Physics* 28, no. 2: 219–47.

———Muller, F. A. (1999). The Equivalence Myth of Quantum Mechanics. *Studies in History and Philosophy of Science Part B: Studies in History and Philosophy of Modern Physics* 30, no. 4: 543–45 (addendum).

Murdoch, D. R. 1987. *Niels Bohr's Philosophy of Physics.* Cambridge and New York: Cambridge University Press.

Nicholson, J. W. 1911. A Structural Theory of the Chemical Elements. *London, Edinburgh, and Dublin Philosophical Magazine and Journal of Science* 22, no. 132: 864–89.

Nordheim, L. 1928. Zur Theorie der thermischen Emission und der Reflexion von Elektronen an Metallen. *Zeitschrift für Physik* 46, no. 11–12: 833–55.

Norton, J. D. 2000. How We Know about Electrons. In *After Popper, Kuhn and Feyerabend,* 67–97. Dordrecht, Netherlands: Springer.

————. 2003. A Material Theory of Induction. *Philosophy of Science* 70, no. 4: 647–70.

————. 2014. Einstein's Special Theory of Relativity and the Problems in the Elecrtodynamics of Moving Bodies That Led Him to It. In *The Cambridge Companion to Einstein*, edited by M. Janssen and C. Lehner. Cambridge: Cambridge University Press.

————. 2018. Einstein on the Completeness of Quantum Theory. Accessed at https://www.pitt.edu/~jdnorton/teaching/HPS_0410/chapters/quantum _theory_completeness/index.html.

Olkhovsky, V. S., E. Recami, and J. Jakiel. 2004. Unified Time Analysis of Photon and Particle Tunnelling. *Physics Reports* 398, no. 3: 133–78.

Orzel C. 2015. The Role of Philosophy in Physics. Forbes, May 11. Accessed at https://www.forbes.com/sites/chadorzel/2015/05/11/the-role-of-philosophy -in-physics/#55c05bf152e0.

Osnaghi, S. 2017. Complementarity as a Route to Inferentialism. In *Niels Bohr and the Philosophy of Physics: Twenty-First-Century Perspectives*, edited by J. Faye and H. J. Folse, 155–78. London and New York: Bloomsbury.

Pais, A. 1968. Reminiscences from the Post-War Years. In *Niels Bohr: His Life and Work as Seen by His Friends and Colleagues,* edited by S. Rozental, 215–26. Amsterdam: North-Holland and New York: Wiley.

————. 1986. *Inward Bound*. Oxford: Oxford University Press.

————. 1991. *Niels Bohr's Times: In Physics, Philosophy, and Polity*. Oxford, UK: Oxford University Press.

Parker, W. S. 2009. Does Matter Really Matter? Computer Simulations, Experiments, and Materiality. *Synthese* 169, no. 3: 483–96.

Pauli, W. 1926. Über die Wasserstoffspektrum vom Standpunkt der nuen Quantenmechanik. In *Sources of Quantum Mechanics*, edited by B. L. Van der Waerdeb, 387–416. Amsterdam: North-Holland, 1967.

————. 1979. *Wissenschaftlicher Briefwechsel mit Bohr, Einstein, Heisenberg u.a., Band I*. Berlin: Springer.

————. 1980. *General Principles of Quantum Theory*. Translated by P. Achuthan and K. Venkatesan. Berlin: Springer Verlag.

Paschen, F. V., and E. Back. 1912. Normale und anomale Zeemaneffekte. *Annalen der Physik* 344, no. 15: 897–932.

Paz, J. P., and W. H. Zurek. 1993. Environment-Induced Decoherence, Classicality, and Consistency of Quantum Histories. *Physical Review D* 48, no. 6: 2728.

Perović, S. 2003. Schrödinger's and Everett's Interpretations of Quantum Mechanics. In *Quantum Theory: Reconsideration of Foundations, 2. Mathematical Modeling in Physics, Engineering and Cognitive Science*, vol. 10, edited by A. Khrennikov. Växjo, Sweden: Växjo University Press, 747–67.

————. 2005. Recent Revival of Schrödinger's Ideas on Interpreting Quantum Mechanics, and the Relevance of Their Early Experimental Critique. *AIP Conference Proceedings* 750, no. 1: 316–20.

————. 2006. Schrödinger's Interpretation of Quantum Mechanics and the

Relevance of Bohr's Experimental Critique. *Studies in History and Philosophy of Science Part B: Studies in History and Philosophy of Modern Physics* 37, no. 2: 275–97.

———. 2008. Why Were Matrix Mechanics and Wave Mechanics Considered Wquivalent? *Studies in History and Philosophy of Science Part B: Studies in History and Philosophy of Modern Physics* 39, no. 2: 444–61.

———. 2013. Emergence of Complementarity and the Baconian Roots of Niels Bohr's Method. *Studies in History and Philosophy of Science Part B: Studies in History and Philosophy of Modern Physics* 44, no. 3: 162–73.

———. 2017. Complementarity and Quantum Tunneling. In *Niels Bohr and the Philosophy of Physics: Twenty-First-Century Perspectives*, edited by J. Faye and H. J. Folse, 207–22. London and New York: Bloomsbury.

———. 2019. *Kvantna revolucija*. Smederevo, Serbia: Heliks.

Perrin, J. 1895. Nouvelles propriétés des rayons cathodiques. *Comptes Rendus* 121:1130–34.

Petersen, A. 1968. *Quantum Physics and the Philosophical Tradition*. Cambridge, MA: MIT Press.

Pickering, A. 1984. Against Putting the Phenomena First: The Discovery of the Weak Neutral Current. *Studies in History and Philosophy of Science Part A* 15, no. 2: 85–117.

Planck, M. 1913. Die Gesetze der Wärmestrahlung und die Hypothese der Elementaren Wirkungsquanten. In *Die Theorie der Strahlung und der Quanten: Verhandlungen auf einer von e. Solvay einberufenen Zusammenkunft* (30 October to 3 November 1911), edited by A. Eucken, 77–94. Berlin: Verlag Chemie.

Platt, J. R. 1964. Strong Inference. *Science*, 146, no. 3642: 347–53.

Plotnitsky, A. 2017. Fragmentation, Multiplicity, and Technology in Quantum Physics: Bohr's Thought from the Twentieth to Twenty-First Century. In *Niels Bohr and the Philosophy of Physics: Twenty-First-Century Perspectives*, edited by J. Faye and H. J. Folse, 179–204. London and New York: Bloomsbury.

Plucker, J. 1858. On the Action of the Magnet upon the Electrical Discharge in Rarefied Gases. *London, Edinburgh and Dublin Philosophical Magazine* 5th series, 44:119–35.

Plucker, J., and J. W. Hittorf. 1865. I. On the Spectra of Ignited Gases and Vapours, with Especial Regard to the Different Spectra of the Same Elementary Gaseous Substance. *Philosophical Transactions of the Royal Society of London* 155:1–29.

Popper, K. 2002. *An Unended Quest.* London and New York: Routledge.

Prizbram, K. 1967. *Letters on Wave Mechanics*. New York: Philosophical Library.

Radder, H. 1982. An Immanent Criticism of Lakatos' Account of the "Degenerating Phase'" of Bohr's Atomic Theory. *Zeitschrift für allgemeine Wissenschaftstheorie* 13, no. 1: 99–109.

Raman, V. V., and P. Forman. 1969. Why Was It Schrödinger Who Developed de Broglie's deas? *Historical Studies in the Physical Sciences* 1:291–314.

Ramsauer, C. 1921. Über den wirkungsquerschnitt der gasmoleküle gegenü ber langsamen elektronen. *Annalen der Physik* 72:345–52.

Rentetzi, M. 2007. *Trafficking Materials and Gendered Experimental Practices: Radium Research in Early 20th Century Vienna*. New York: Columbia University Press.

Ritz, W. 1908. On a New Law of Series Spectra. *Astrophysical Journal* 28:237.

Rosenfeld, L. 1968. Niels Bohr in the Thirties: Consolidation and Extension of the Conception of Complementarity. In *Niels Bohr: His Life and Work as Seen by His Friends and Colleagues,* edited by S. Rozental, 114–36. Amsterdam: North-Holland, and New York: Wiley.

Rozental, S. 1968. *Niels Bohr: His Life and Work as Seen by His Friends and Colleagues*. Amsterdam, North-Holland, and New York: Wiley.

Rutherford, E. 1962. *The Collected Papers of Lord Rutherford of Nelson*. New Zealand, Cambridge, and Montreal: Allen & Unwin.

Rutherford, E., and H. Geiger. 1908a. An Electrical Method of Counting the Number of α-Particles from Radio-Active Substances. *Proceedings of the Royal Society of London. Series A, Containing Papers of a Mathematical and Physical Character* 81, no. 546: 141–61.

———. 1908b. The Charge and Nature of the α-Particle. *Proceedings of the Royal Society of London. Series A, Containing Papers of a Mathematical and Physical Character* 81, no. 546: 162–73.

Rydberg, J. R. 1890. Recherches sur la constitution des spectres d'émission des éléments chimiques. *Kungliga vetenskapsakademiens handlingar* 23, no. 11.

Sargent, R-M. 2001. Baconian Experimentalism: Comments on McMullin's History of the Philosophy of Science. *Philosophy of Science* 68, no. 3: 311–18.

Scheibe, E. 1973. *The Logical Analysis of Quantum Mechanics*. Oxford: Pergamon Press.

Schlosshauer, M. 2004. Decoherence, the Measurement Problem, and Interpretations of Quantum Mechanics. *Reviews of Modern Physics* 76:1267–1305.

Schickore, J. 2011. More Thoughts on HPS: Another 20 Years Later. *Perspectives on Science* 19, no. 4: 453–81.

Schindler, S., and R. Scholl. forthcoming. Historical Case Studies: The "Model Organisms" of Philosophy of Science. *Erkenntnis*. Accessed at https://doi.org/10.1007/s10670-020-00224-5.

Schlick, M. 1979. Quantum Theory and the Knowability of Nature. *Philosophical Papers* 2:482–90.

Schulte, O. 2002. Formal Learning Theory, *Stanford Encyclopedia of Philosophy*. Accessed at https://plato.stanford.edu/entries/learning-formal/#Aca.

Schrödinger, E. 1926a. Quantisation as a Problem of Proper Values, I. In *Collected Papers on Wave Mechanics*, 1–12. New York: Chelsea. Original work published in *Annalen der Physik* 79.

———. 1926b. Quantisation as a Problem of Proper Values, II. In *Collected Papers on Wave Mechanics*, 13–40. New York: Chelsea. Original work published in *Annalen der Physik* 79.

———. 1926c. Quantisation as a Problem of Proper Values, III. In *Collected Papers on Wave Mechanics*, 62–101. New York: Chelsea. Original work published in *Annalen der Physik* 80.

———. 1926d. Quantisation and Proper Values, IV. In *Collected Papers on Wave Mechanics*, 102–23. New York: Chelsea. Original work published in *Annalen der Physik* 81.

———. 1926e. On the Relation between the Quantum Mechanics of Heisenberg, Born, and Jordan and that of Schrödinger. In *Collected Papers on Wave Mechanics*, 45–61. New York: Chelsea. Original work published in *Annalen der Physik* 79.

———. 1927a. The Compton Effect. In *Collected Papers on Wave Mechanics*, 124–29. New York: Chelsea. Original work published in *Annalen der Physik* 82.

———. 1927b. The Energy-Momentum Theorem for Material Waves. *The Collected Papers on Wave Mechanics*, 130–36. New York: Chelsea. Original work published in *Annalen der Physik* 82.

———. 1930. Zum Heisenbergschen Unschärfeprinzip. *Sitzungsberichte der Preussischen Akademie der Wissenschaften, Physikalisch-mathematische Klasse* 14:296–303.

———. 1956. Are There Quantum Jumps? In *What is Life? and Other Scientific Essays*. Garden City, NY: Doubleday Anchor.

Scott, W. T. 1967. *Erwin Schrödinger: An Introduction to His Writings*. Amherst, MA: University of Massachusetts Press.

Sebens, T. C. 2020. What's Everything Made of? *American Scientist* 108, no. 1: 42–50.

Serwer, D. 1977. Unmechanischer Zwang: Pauli, Heisenberg, and the Rejection of the Mechanical Atom, 1923–1925. *Historical Studies in the Physical Sciences* 8:189–256.

Seth, S. 2010. Crafting the Quantum: Arnold Sommerfeld and the Practice of Theory, 1890–1926. Cambridge, MA: MIT Press.

Sommerfeld, A. 1921. *Atombau und Spektrallinien.* 2nd edition. Braunschweig, Germany: Vieweg & Sohn.

———. 1923. *Atomic Structure and Spectral Lines.* London: Methuen.

Sparenberg, J. M., and D. Gaspard. 2018. Decoherence and Determinism in a One-Dimensional Cloud-Chamber Model. *Foundations of Physics* 48, no. 4: 429–39.

Staley, R. 2008. Worldviews and Physicists' Experience of Disciplinary Change: On the Uses of "Classical" Physics. *Studies in History and Philosophy of Science Part A* 39, no. 3: 298–311.

———. 2018. Sensory Studies, or When Physics Was Psychophysics: Ernst Mach and Physics between Physiology and Psychology, 1860–71. *History of Science.* Accessed at https://doi.org/10.1177/0073275318784104.

Stapp, H. P. 1972. The Copenhagen Interpretation. *American Journal of Physics* 40, no. 8: 1098–1116.

Stegenga, J. 2009. Robustness, Discordance, and Relevance. *Philosophy of Science* 76, no. 5: 650–61.

Stuewer, R. 1975. *The Compton Effect*. New York: Science History Publications.

Thirring, H. 1928. Die Grundgedanken der neueren Quantentheorie. In *Ergebnisse der exakten naturwissenschaften*, 384–431. Berlin and Heidelberg: Springer.

Thomson, J. J. 1897. On Cathode Rays. *Philosophical Magazine* series 5, 44:293–316.

———. 1903. *Conduction of Electricity through Gases*. Cambridge: Cambridge University Press.

———. 1904. *Electricity and Matter*. New Haven: Yale University Press.

———. 1970. *The Royal Institution Library of Science*, vol. 5, edited by W. Bragg and G. Porter, Amsterdam: Elsevier.

Van Fraassen, B. C. 2008. *Scientific Representation: Paradoxes of Perspective*. Oxford: Oxford University Press.

———. 2009. The Perils of Perrin, in the Hands of Philosophers. *Philosophical Studies* 143, no. 1: 5–24.

Vickers, P. 2014. Scientific Theory Eliminativism. *Erkenntnis* 79, no. 1: 111–26.

Von Laue, M. 1934. Uber Heisenbergs Ungenauigkeitsbeziehungen und ihre erkenntnistheoretische Bedeutung. *Naturwissenschaften* 22, no. 26: 439–41.

Von Neumann, J. 1932. *Mathematical Foundations of Quantum Mechanics*, translated by R. T. Beyer. Princeton, NJ: Princeton University Press.

Weinert, F. 2001. The Construction of Atom Models: Eliminative Inductivism and Its Relation to Falsificationism. *Foundations of Science* 5:491–531.

Wessels, L. 1979. Schrödinger's Route to Wave Mechanics. *Studies in History and Philosophy of Science Part A* 10, no. 4: 311–40.

Wessels, L. 1983. Erwin Schrödinger and the Descriptive Tradition. *Springs of Scientific Creativity* 272:284–91.

Whittaker, E. 1953. *Aether and Electricity, Vol. II: The Modern Theories*. New York: Harper and Brothers.

Wien, W. 1926. *Vergagenheit, Gegenwart und Zukunft der Physik: Rede Gehalten beim Stiftungsfest der Universität München am 19 Juni*. Munich: Max Hueber Verlag.

Winful, H. G. 2006. Tunneling Time, the Hartman Effect, and Superluminality: A Proposed Resolution of an Old Paradox. *Physics Reports* 436, no. 1: 1–69.

Wolff, J. 2014. Heisenberg's Observability Principle. *Studies in History and Philosophy of Science Part B: Studies in History and Philosophy of Modern Physics* 45:19–26.

Woodward, J. 2006. Some Varieties of Robustness. *Journal of Economic Methodology* 13, no. 2: 219–40.

Worral, J. 2000. The Scope, Limits and Distinctiveness of the Method of "Deduction of Phenomena": Some Lessons from Newton's "Demonstrations" in Optics. *British Journal for Philosophy of Science* 51:45–80.

Zeh, H. D. 1970. On the Interpretation of Measurement in Quantum Theory. *Founations of Physics* 1:69–76.

Zinkernagel, H. 2016. Niels Bohr on the Wave Function and the Classical/Quantum Divide. *Studies in History and Philosophy of Science Part B: Studies in History and Philosophy of Modern Physics* 53:9–19.

Zurek, W. H. 1981. Pointer Basis of Quantum Apparatus: Into What Mixture Does the Wave Packet Collapse? *Physical Review D* 24:1516–25.

INDEX